Radiation Physics Applications

Radiation Physics Applications

Krina Shah

ARCLER
PRESS

www.arclerpress.com

Radiation Physics Applications

Krina Shah

Arcler Press
4164 Lakeshore Road
Burlington, ON L7L 1A4
Canada
www.arclerpress.com
Email: orders@arclereducation.com

ISBN: 978-1-77956-969-1 (Hardcover)

Arcler Press publishes wide variety of books and eBooks. For more information about Arcler Press and its products, visit our website at www.arclerpress.com

ABOUT THE AUTHOR

Krina Shah is a dedicated mathematics tutor with an extensive background in algebra and calculus. She is currently employed in a tutoring role where she rigorously analyzes word problems, interprets data reports, and engages in research activities. Krina holds a Master's degree from the State University of Gujarat and is deeply committed to continuous learning. Her dedication to professional development is reflected in her active participation in various extracurricular activities. In addition to her professional commitments, Krina devotes her time to volunteering with organizations, including blood donation camps, and teaching children with disabilities. Her passion for mathematics, combined with her broad range of experiences, contributes to her development as a compassionate and skilled educational professional.

TABLE OF CONTENTS

LIST OF FIGURES

PREFACE

The manufacture of metallic materials for a range of applications that meet the needs of economic efficiency and minimal ecological disturbance greatly influences the development of energy based on nuclear power engineering, as well as future advancements in space exploration. The use of metals and alloys with strong radiation resistance and rapid reduction of radioactivity is crucial for the successful realization of these objectives. The advancement of the concepts pertaining to the mechanisms governing modifications in the structure, characteristics, and activation parameters of irradiated objects, as well as the analysis of the mechanisms and the creation of strategies to lessen the harmful effects of radiation, form the basis of the physical foundations and principles guiding the development of these materials. The design and operation of nuclear reactors, nuclear power plants, nuclear fuel cycle facilities, and the decontamination and decommissioning of these facilities at the end of their span are all included in the field of nuclear physics and engineering. Atomic and radiation physics are used by plasma physicists to interpret observations in a variety of fields, including magnetic fusion and low-temperature plasmas with technological significance. With the building of the ITER tokamak, new laser-plasma facilities, and the creation of novel techniques for producing plasma, plasma physics is expanding. The theoretical creation and simulation of laboratory and astrophysical plasmas heavily relies on atomic and radiation physics. The Manhattan Project resulted in the creation and application of the atomic bomb and served as the foundation for the US nuclear industry. It still has a negative impact on the nuclear sector today. The Manhattan Project aimed to produce pure plutonium-239 contaminated with other plutonium isotopes and highly enriched uranium. These materials are currently being diluted in excess so that nuclear-powered electric generating plants can use them. The commercial nuclear power plant is viewed by many as a threat to both the environment and human life. This is linked with the nuclear reactor's history of atomic weapons and in part with the reactor accidents that happened in 1979 at the Three Mile Island nuclear power plant in Harrisburg, Pennsylvania, and in 1986 at the Chernobyl nuclear power plant near Kiev, Ukraine.

This book helps the students of all grades who want to know the basics of nuclear physics.

INTRODUCTION

In a nuclear power plant, the nuclear reactor serves as a heat source for the steam that powers an electric generator turbine. It is identical to burning coal or natural gas in a boiler. The distinction is that energy is produced by splitting an atom rather than by burning fossil fuels. One gram of uranium can provide 1 MW day of energy when split or fissioned, demonstrating the importance of a concentrated energy source. The yield of one gram of coal is less than 0.01 kW hours. The United States started building nuclear power plants in the 1950s. The first power plant in the US to start up was in Pennsylvania. A number of pilot plants of different designs, the majority of which had power producing capacities under 100 MW, came after that. There was a rush to construct larger nuclear-powered power plants in the late 1960s. The United States stopped building nuclear power reactors after the Three Mile Island accident. Today's nuclear sector is global in reach. Large design and manufacturing firms operate globally. In order to fulfill the upcoming demands for the production of electric energy, the US industry is recovering after surviving the 30 years after the Three Mile Island tragedy. Nuclear power saw a resurgence in 2008. This shift was caused by the cost-effectiveness of constructing new nuclear power facilities in comparison to huge fossil fuel plants, as well as uncertainties about reducing emissions from the latter. China and India are witnessing significant increases in nuclear power, whereas slower economic growth and the availability of natural gas as an electricity-generating fuel are restraints on growth in other regions. There is a greater chance of nuclear incidents and accidents when nuclear power technology is introduced to less developed countries with less operational and technical expertise. Even developed countries can experience nuclear accidents, as demonstrated by the 2011 Fukushima Daiichi accident. Nuclear power facility licensing requirements need to be closely studied to make sure they can maintain their fission product barriers in the event of both natural and man-made disasters. Radioactive material can be released into the environment more easily as fission product barriers degrade, which might have serious negative effects on the environment and the economy. Accidents at power reactors are not the only way that radioactive material leaks can damage human life and the environment. The nuclear methods and materials are improving medical imaging and treatment processes. The effectiveness of therapies is improved and adverse side effects are reduced by new methods that deliver a tailored dose to the tumor site while decreasing the dose absorbed by healthy tissue. After therapy, these methods ought to improve a patient's quality of life.

CHAPTER 1

INTRODUCTION TO RADIATION PHYSICS

Contents

1.1. INTRODUCTION

Nuclear radiation is prevalent, and it exists in the atmosphere as cosmic rays originating from outer space, the sources of which remain unidentified (Cerrito, 2017). Carbon, which possesses a minor but finite radioactive isotope, is also found in plants; this characteristic is utilized to effectively ascertain the age of archeological samples. Most organisms, including humans, contain calcium, which comprises a small quantity of the radioactive isotope potassium-40

(40K) (Abdelaleem, 2017). During the 20th century, researchers achieved considerable advancements in utilizing energy from nuclear fission, applying nuclear methodologies for non-destructive material testing, as well as medical diagnostics and treatment. Exponential growth and decay are prevalent in the physical sciences. In the context of growth, we can delineate a maximum threshold to which a sample, when left undisturbed, will expand or grow after an indefinite period. In the event of decay, a sample will disappear only after an indefinite period. Both growth and decay will be minimal after a finite duration. The mean life in radioactive decays is a characteristic time unique to the decaying nucleus. It is independent of the physical or chemical state (material). Exponential processes possess a significant characteristic whereby the decay or growth rate inside any period is based on the quantity present at that interval, rather than its historical context. The decay rate (activity) at any given moment is contingent upon the quantity of radioactive nuclei present in the sample at that time, rather than the time of its preparation in the past. By utilizing data on the half-life or mean life of a radioactive element, one can determine the quantity of radioactive atoms in a sample, provided the sample activity is known, or vice versa. We can subsequently forecast its activity or the quantity of radioactive atoms in the sample at any future point and delineate its history. The activity of a radioactive sample equals the product of the quantity of radioactive atoms and the decay constant. A higher mean life corresponds to a smaller decay constant, resulting in a sample that is less radioactive than an equivalent quantity of material with a shorter mean life or a larger decay constant. Researchers introduced units of radioactivity to quantify the measurements. The International System of Units designates the unit Becquerel, abbreviated as Bq. A sample is classified as having a radioactivity level of 1 Bq if the disintegration rate is one decay per second. Long-lived activities are important when radioactive nuclei are generated in substantial quantities. A reduced number of atoms is required to generate substantial quantities of short-lived activity.

1.2. RADIOACTIVITY

For several applications, it is essential to process the materials, transport them to the application site, and conduct the measurements. Consequently, one frequently requires activities with finite, rather lengthy half-lives. It is essential to differentiate between the decay rates and radiation levels of nuclear decay (Figure 1.1). The quantity of radiation emitted during nuclear decay is specific to the emitting nucleus or the nuclear activity being examined. Although the decay rates of two sources may be identical, the radiation levels will be elevated for the process that produces a greater quantity of radiation per decay. A 137Cs nucleus emits a beta particle and a 661 keV photon on each decay event (D'Auria,

2018). A 60Co nucleus generates a beta particle and two photons with energies of 1,173 and 1,333 keV per decay, resulting in higher radiation levels for 60Co compared to 137Cs at equivalent decay rates (Golovko, 2022). 152Eu emits multiple gamma rays per decay, and its radiation levels significantly exceed those of 137Cs and 60Co with equivalent decay rates (Lépy et al., 2006).

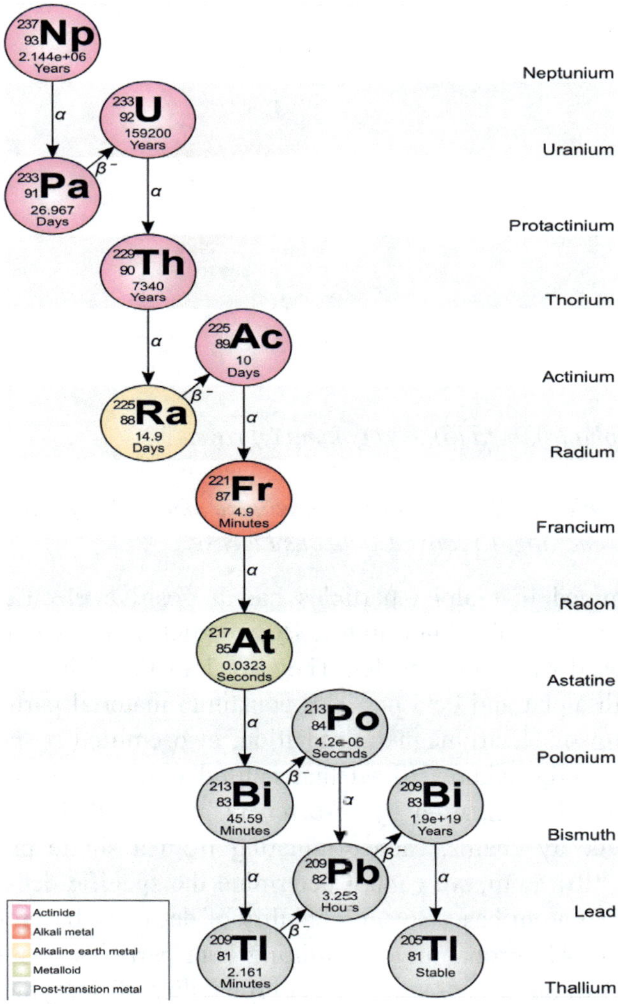

Figure 1.1. *Nuclear decay.*

Source: Audi, G., Bersillon, O., Blachot, J., & Wapstra, A. H. (2003). The Nubase evaluation of nuclear and decay properties. Nuclear Physics A, 729(1), 3–128. https://doi.org/10.1016/j.nuclphysa.2003.11.001, Creative Commons License.

Radioactivity was initially identified in natural samples. Becquerel's finding that an exposed photographic film was affected by emissions from a uranium sample was significant. Within a span of five years, three categories of radiation were identified. They were designated as alpha (α), beta (β), and gamma (γ) rays (Figure 1.2).

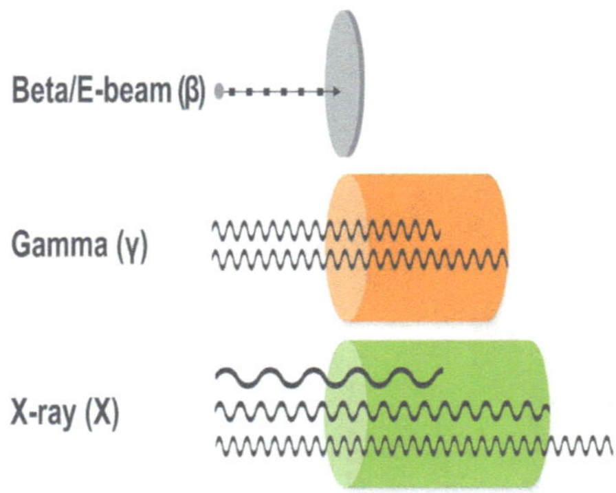

Figure 1.2. *Alpha (α), beta (β), and gamma (γ) rays.*

Source: https://upload.wikimedia.org/wikipedia/commons/2/24/Beta-e-beam-Gamma-X-ray-efficiency.jpg, Creative Commons License.

It was determined that alpha particles had a positive electric charge and constitute the nuclei of helium atoms. Beta particles are negatively charged particles released by atomic nuclei. They are indistinguishable from atomic electrons. Both alpha and beta particles constitute material particles. Gamma rays are a form of electromagnetic radiation, also emitted by atomic nuclei. It was subsequently acknowledged that natural radioactivity is an intricate event. Extensive investigation by researchers has revealed the existence of three distinct decay chains, each originating from a single progenitor. For an individual ^{212}Bi atom, we cannot determine the specific decay pathway it will take. A ^{212}Bi atom has a 36% probability of decaying by producing alpha particles and a 64% probability of emitting beta particles (Garashchenko et al., 2018). A ^{212}Bi atom possesses a 36% probability of decaying via the first pathway and a 64% probability of decaying via the second pathway (Hassfjell and Brechbiel, 2001). Natural uranium ore samples exhibit gamma rays with energy that may be associated with members of the 4n+2 series, specifically ^{238}U. The identifications are derived from the energies and intensities of gamma-ray emissions from the ore sample, in comparison to data available in databases, which compile analyses from various research studies. In sequential decays, a radioactive atom undergoes decay to provide a product with its specific half-life. The unstable decay product subsequently decays into the next element in the series, each possessing its own distinct half-life, and continues in this manner. In the terminology of radiation, we designate them as parent, daughter, granddaughter, and so on. In a sample, all daughter products are reported to be in secular equilibrium with the parent activity (^{238}U), and the emission intensity variations adhere to the half-life of ^{238}U (Figure 1.3).

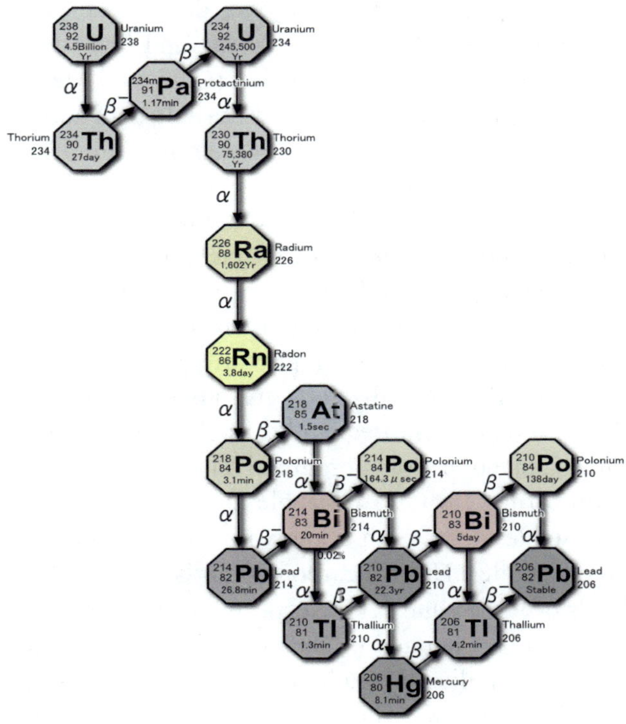

Figure 1.3. *Half-life of ^{238}U.*

Source: https://upload.wikimedia.org/wikipedia/commons/a/a1/Decay_chain%284n%2B2%2C_Uranium_series%29.PNG, Creative Commons License.

In a series of sequential decays, the relative lifetimes of progenitors and descendants affect the growth and decay rates of the descendants. There are two recognized forms of radioactive equilibria: secular and transient equilibrium. In a sequential decay, a daughter nuclide is in secular equilibrium with its parent if the emissions of the daughter product correspond to the mean lifetime of the parent nuclide. Only a limited number of naturally occurring radiation exist. Artificially generated radioactive compounds are utilized in laboratories. An essential criterion is the inexpensive and effective distribution of optimal quantities of radioactivity to clients or stakeholders. Isotope synthesis can be achieved by bombarding a stable isotope with neutrons from a nuclear reactor or with particle beams from a particle accelerator. This is applicable to all production processes. We subject a target material to a particle beam, yielding a steady production rate R of radioactive atoms with a disintegration constant λ per second. Throughout the production period, the quantity of radioactive atoms increases over time t, concurrently undergoing decay as they are generated. The age estimation of archaeological, geological, and forensic samples relies on the fundamental radioactive decay equation (Cook and MacKenzie, 2014). In a sample containing radioactive atoms, when left undisturbed, the rate of disintegration and the quantity of radioactive atoms diminish exponentially with time. By measuring the specific activity or the quantity of atoms of a particular species, we can ascertain the age of a sample. If A_0 and N_0 represent the particular activity and the quantity of unstable atoms of the sample at its inception, then the activity A(t) and N(t) after a time interval t are expressed as Living biological specimens, including organisms and plants, continuously exchange carbon with their environment, primarily as carbon dioxide. Humans exhale carbon dioxide, which is utilized by green plants. On death, the samples cease this carbon exchange. Subsequently, as the quantity of stable atoms remains constant, radioactive atoms progressively diminish according to their half-lives. By comparing the ratio of stable atoms to either radioactivity or the quantity of unstable atoms with a known disintegration constant, we can determine the age (sample). Natural carbon comprises two stable isotopes, ^{12}C and ^{13}C, with loads of 98.9% and 1.1%, respectively. Current C contains traces of unstable ^{14}C, which is generated in the atmosphere by the interaction of air molecules with cosmic radiation, including photons, neutrons, and charged particles. The atomic weight of carbon can be attributed to ^{12}C atoms.

The half-life of ^{14}C ($t_{1/2}$ = 5730 years) is short relative to geological time periods spanning several billion years. Earth scientists utilize decay chains, each with its specific applicability, to ascertain the ages of rock formations (Clauser, 2024). A query on the NNDC[13] website reveals the existence of 27 decay chains with half-lives exceeding 1 million years. The isotopes ^{238}U, ^{235}U, ^{232}Th, ^{129}I, ^{40}K, and ^{87}Rb are mainly utilized for dating rocks and minerals. Astrophysicists utilize the decay of ^{26}Al as a chronometer for chondrules, which is crucial for comprehending our planetary system. In the 1970s, the direct counting of radioactive atoms was suggested as a method to improve the sensitivity of small samples. 10 billion atoms of ^{14}C yield 0.2 Bq (Bair et al., 1981). The disintegration rate equals the product of the quantity of radioactive atoms in the sample and the disintegration constant (λ). The samples with extended mean lifetimes necessitate a greater quantity of radioactive atoms to attain equivalent decay rates. Counting radioactive atoms, rather than decays, may yield enhanced precision in rapid measurements. This constituted the concept underlying accelerator mass spectrometry. We need to examine why certain nuclei undergo disintegration while others remain stable. Nuclei disintegrate by producing certain types of radiation. A ^{14}C nucleus, possessing a half-life of 5730 years, undergoes decay through the emission of a beta particle, while ceasing from producing an alpha particle, proton, or any other particle (Belyaev and Ross, 2021). Gamma rays are not released. In general, daughter nuclei generated during nuclear decays emit gamma rays; nevertheless, they do not release neutrons, protons, or other complex nuclei. Physicists have been generating various nuclei using artificial methods utilizing particle accelerators and nuclear reactors. A system, when isolated, will naturally go towards its lowest energy state, dissipating surplus energy as kinetic energy and radiation. A system may consist of nuclei, particles, and/or radiation, or other combinations. In the energy balance, we do not differentiate between material particles and photons. We generally mention the rest masses and kinetic energy of particles. The essential determining factor for physical phenomena is the total energy, which is the aggregate of energies from rest masses and the kinetic energies of particles. An antineutrino is ejected alongside an electron in beta decay, while a neutrino is released during decays involving electron capture or positron emission. Due to the insignificant mass of a neutrino, its presence can be disregarded in the calculation of Q-values for these decays. Both positron emission (β^+ decays) and electron capture yield the same ultimate nucleus. The distinction lies in the fact that β^+ decay emits a positron and a neutrino, whereas electron capture does not include the emission of a positron. No charged particle is released during electron capture. Every atomic nucleus possesses distinct excitation energy levels. Frequently, radioactivity through the emission of an alpha or beta particle leaves the daughter nucleus in an excited state. Gamma ray emissions would not occur if only ground states are occupied. The energetics will ascertain the highest excitation energy obtained during radioactive decay.

1.3. ENERGY

Energy measurements of radiations have been employed as potent spectroscopic instruments to ascertain the actual energy of excited states of daughter nuclei. The association is direct for charged particle decays, as there exists a one-to-one link between charged particle energy and excitation energy. The analysis of gamma ray spectra is not easy, as a cascade of gamma rays may link two levels through intermediary excitations. Information can be inferred from the Q-value of decay and the kinetic energy of alpha particles. We can derive insights from experimentation and articulate the findings. In instances of uncomplicated decays, such as the emission of a single gamma ray from ^{137}Cs or two gamma rays from ^{60}Co, the experiments and interpretations are simple (Jarrett, 1982). Complex decays involving many gamma ray energies necessitate intricate experiments, and the interpretations may be challenging. In nuclear decays, if the mass of the original entity exceeds the total mass of the resultant entities, Q is positive. Energetically, the starting body may transform into the final products, with the surplus energy released as kinetic energy, radiation, or other energy forms. We can ascertain the kinetic energy of products entering detectors or traversing a medium. In decays resulting in two entities in the final state, each body possesses a specified kinetic energy. The kinetic energy of the daughter nucleus is almost zero. It is important that in alpha decays that populate a level in a daughter nucleus, all alpha particles possess a distinct energy. They possess a single energy level. In beta decays that populate a single level, the beta particles exhibit a range of energies rather than a single energy value. They possess a continuum of energies ranging from zero to a maximum value equivalent to the Q value. Utilizing established Q-values, we can ascertain the kinetic energies of particles and conduct additional experiments; from experimentally measured kinetic energies, we can infer information regarding nuclear levels. During the decay, a neutrino with no electric charge and an almost negligible mass is emitted. An energetic system can be established by hitting a target body with energetic photons or other particle beams, including protons, neutrons, and alpha particles. Current advancements have made available particle beams comprising various nuclear species, unstable particles, photons, and electrons across diverse energy ranges. Nuclei, when energized beyond a specific threshold, will undergo disintegration into lighter nuclei, accompanied by the emission of a neutron, proton, alpha particle, etc. The excitation energy of a nucleus at which these emissions become energetically feasible is referred to as the separation energy. Each species possesses a specific energy, including neutron separation energy, proton separation energy, and alpha separation energy, corresponding to the emission of a neutron, proton, or alpha particle, respectively. Understanding these energies is highly beneficial as it can elucidate which energies render a given nucleus unstable to the emission of neutrons, protons, and other particles. A medical physicist evaluates the impact of radiation on living cells during its application in diagnostic imaging or therapeutic interventions. A medical physicist can analyze radiation – interaction effects for the purpose of medical image reconstruction. A physicist assesses radiation doses or determines radiation shielding requirements for occupational safety. A scientist or engineer engages in the design of radiation facilities, including accelerators, reactors, or beamlines for radiation delivery. Environmental scientists engage in assessing the impacts of radiation toxicity on natural ecosystems. Space scientists or engineers engage in studies involving satellites or a space station situated in extreme cosmic radiation settings. Charged particles include electrons, protons, muons, pions, alpha particles, heavy ions, and others. Medical physicists and radiation safety scientists encounter all particles except muons. Muons may also be utilized in the treatment of cancer and AIDS (Scheck, 1978). Neutral radiations include neutrons, neutrinos, and photons. There exists a basic distinction in the energy dissipation mechanisms of electrically charged particles relative to those of neutral radiations. Charged particles have continuous collisions during their paths, resulting in a gradual loss of energy. Neutral radiations experience random collisions in a discontinuous manner, resulting in non-continuous energy loss. For charged particles, we can define energy loss per unit distance traveled and range, which refers to the distance they traverse within a medium. For neutral radiations, we can define the attenuation of intensity per unit distance traversed.

Figure 1.4. *CERN.*

Source: https://upload.wikimedia.org/wikipedia/commons/4/46/Views_of_the_LHC_tunnel_sector_3-4%2C_tirage_2.jpg, Creative Commons License.

It is denoted by an attenuation coefficient, which quantifies the medium efficiency in eliminating radiation from the primary particle beam. All these characteristics are based on the type of radiation, its energy, and the qualities of the medium. The energy loss mechanisms of heavy particles are simple for the majority of practical energy levels. As they traverse, they dissipate energy, experiencing elastic and inelastic collisions with atoms and molecules in the medium, resulting in ionization. The charged particles excluding electrons are considered heavy particles. At low momenta, the energy loss follows a specific curve for each particle type, differentiated by their mass factor. The lighter particles diminish at lower momenta. This indicates that lowest ionization occurs for pions and muons at energies exceeding 0.3 GeV/c, and for protons at around 1.3 GeV/c (Cook et al., 2017). Above the minimal ionizing zone, energy losses are largely stable to particle mass. The energy loss measurements are not the effective means of identifying particles at really high energies. In recent years, CERN (Figure 1.4) researchers have demonstrated that precise measurements at extremely high energies could aid in particle identification, given the energy loss plateau varies in magnitude for each particle. A defining aspect of charged particle propagation is the largest energy loss over a short distance, which occurs at the end of the particle trajectory or trail, referred to as the Bragg peak.

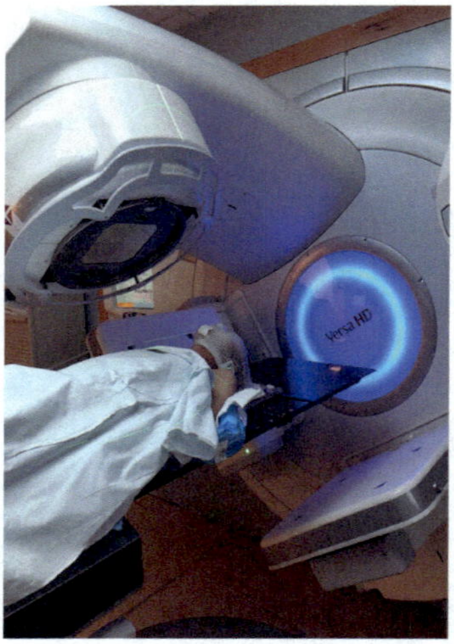

Figure 1.5. *Radiation therapy.*

Source: https://upload.wikimedia.org/wikipedia/commons/4/44/Radiation_therapy_for_cancer.jpg, Creative Commons License.

This property is extensively utilized in radiation therapy. At energies within the threshold of minimal ionization, energy loss escalates as the particle energy diminishes. The energy dissipation per unit length escalates when the particle decelerates within the denser medium. As an ion expends energy, the rate of energy loss diminishes and ultimately reaches zero. A particle cannot lose more than its own kinetic energy. A particle, while moving through a material, undergoes increasing energy loss internally, succeeded by diminishing energy until it ceases motion. This phenomenon is known as the Bragg Peak. As it approaches the ending of its trajectory, close to the range at which it ceases motion, a particle will possess minimal energy to dissipate. In certain interactions, it may possess less energy than the thermal energies of its colliding counterparts and may acquire some energy. This will result in particle straggling, meaning that not all particles of same energy cease at the same depth within the medium, leading to a narrow but finite range throughout which the particles stop. The range and depth of the Bragg peak are based on the particle type, its energy, and the medium through which it propagates.

1.4. APPLICATION

Radiation therapy (Figure 1.5) for cancers utilizing ions, such as protons or carbon ions, leverages this to optimize beam energies, hence maximizing energy deposition within the tumor (Kraft, 2000). The charged particles possess a limited range, with the most energy being deposited at the end of that range. The characteristic of heavy ions is advantageous for radiation therapy as it decreases the dosage to adjacent, non-targeted tissues. Energy loss is minimal at reduced thicknesses, when alpha particles possess high energies. In the deeper strata of the medium, where alpha particles possess diminished energy, the energy loss is substantial. The existence of a Bragg peak (Figure 1.6) at the end of the alpha particle trajectory is clear. This characteristic is utilized in radiation therapy employing ion beams.

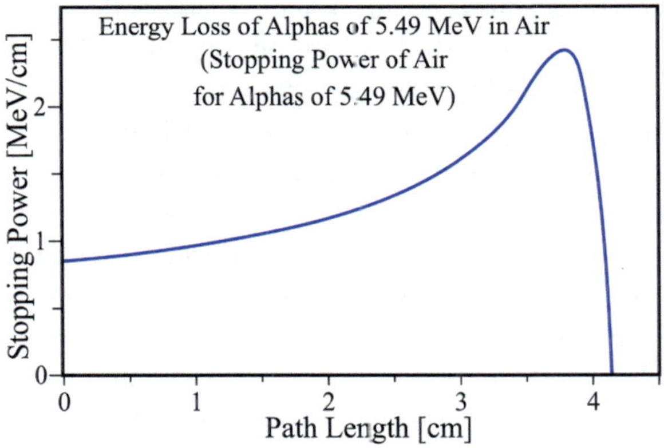

Figure 1.6. *Braggs peak.*

Source:https://upload.wikimedia.org/wikipedia/commons/f/f9/Bragg_Curve_for_Alphas_in_Air-PT-en.svg, Creative Commons License.

Bragg's additivity rule neglects the influence of electron binding effects in chemical compounds. Many researchers examined the constraints of Bragg's rule. There is consensus that Bragg's rule has minimal validity at extremely low energies and may deviate by approximately 50% from measurements, particularly at levels where electron binding effects are significant. The scattering effects of a particle are within the medium, as well as the straggling is caused by energy fluctuations experienced by the particles. This phenomenon occurs when particles decelerate to minimal energy levels, resulting in a tendency to either lose or acquire energy from interacting entities. These occurrences result in particles remaining along their trajectory or deviating from their primary trajectories. Electrons, as they ionize media, interact with atomic electrons, which are their counterparts. Their interactions with the electrons of material media differ from those of massive particles. As electrons decelerate, interactions with atomic electrons in the medium can lead to significantly larger deflections, causing them to be randomly diverted off their trajectory. Due to their small mass, they attain velocities approaching that of light at minimal energy levels and can dissipate energy through the emission of electromagnetic radiation. The total energy loss of electrons results from medium ionization and radiation emission. The interactions between photons and electrons with matter are interconnected.

1.5. INTERACTIONS

The interaction of photons with matter results in the emission of electrons and the creation of electron-positron pairs. Like other charged particles, electrons experience continual deceleration due to collisions with atomic electrons in material media. They dissipate energy by radiation emission, which then propagates in a distinctive manner. Charged particles traverse a medium, creating ionization trails along their trajectory. It can be regarded as a continuous deterministic process for all practical purposes, while it is regulated by quantum mechanics statistical processes at a microscopic level. A photon can traverse an unspecified distance without engaging with other contact or body. Upon interaction, it is detached from the beam, leading to a reduction in beam intensity. We can therefore denote the reduction in the intensity of a photon beam as a result of interactions inside a medium, rather than as energy loss. The likelihood of a photon interacting inside a length interval dx of a medium is unaffected by the distance it has traversed prior to reaching the area of interest. This feature indicates an exponential reduction in photon beam intensity. We can also observe the resemblance to radioactive decay. The decay probability is unaffected by the past of a radioactive atom. Atoms or molecules display several excited states. They can be stimulated from the ground state to specific levels when incident photons possess the appropriate energy. Upon exposure to external radiation, photons that match the excitation energy of atoms or molecules are preferentially absorbed, a phenomenon referred to as resonant absorption. The sudden fluctuations in the attenuation coefficients result from an increase in photon absorption when a photon energy aligns with these excitations. By positioning a photon detector away from the primary beam trajectory, we may identify the characteristic spectral lines of atoms. This property is extensively utilized in scientific study and other applications for the identification of material structures using procedures referred to as resonance fluorescence. The energy loss mechanism of high-energy electrons in high Z medium is intricately connected to that of photons, as these electrons emit photons throughout their traversal (Yuan and Liu, 2022). Consequently, the interactions of photons, electrons, and positrons exhibit similarities in their underlying physical properties. The energy dissipation of these organisms results from electromagnetic interactions. Low energy electrons and positrons ionize medium while emitting radiation at higher energies. For charged particles, we consider elastic scattering and ionization as the primary mechanisms of energy loss, which occurs continuously. Although electrons ionize continuously, radiative processes predominantly account for the energy losses of high-energy electrons, and these processes are discontinuous. In radiation, energy loss during a single interaction may occur with or without the presence of a photon in the end state. A single interaction with a partner may lead to the formation of a charged particle-antiparticle pair, which will then dissipate their energy. At higher energies, the interactions among photons, electrons, and positrons are not clear. In nuclear and radiation physics, the observed photon energies correspond to electron-positron pairs. At energies exceeding several 100 MeV, it is anticipated that pairs of heavier particle-antiparticle combinations, such as pions or much heavier counterparts, can be generated, provided that the criteria for energy and momentum conservation are met. Since the inception of cosmic radiation research, it has been established that cosmic radiation permeates the whole surface of the Earth and consists of photons and electrons, among other components. A physicist named Homi Bhabha (Figure 1.7) identified this phenomena as shower development, now referred to as Bhabha showers (Bhabha et al., 1997). While interacting with matter, it may dislodge an electron from an atom or molecule, giving its energy to the electron. The photon may generate particle-antiparticle pairs. If they are electron-positron pairs of adequate energy, they may produce radiation termed bremsstrahlung and then travel with diminished energy in an alternate direction. Positrons can annihilate with electrons in the medium, resulting in the emission of a pair of photons. The interaction processes of photons, although constant functions of the medium atomic number, exhibit variation with respect to photon energy and atomic number in a specific manner. This capability is utilized to our benefit for each application of interest. Rayleigh scattering (Figure 1.8) refers to the elastic scattering of photons by atoms or molecules within the medium. It is coherent scattering, in which the atom undergoes recoil. It entails neither excitation nor ionization of the atom. This mechanism leads to a negligible loss of photon energy; it is significant at extremely low photon energies and contributes minimally to overall photon energy loss.

Figure 1.7. *Homi Bhabha.*

Source:https://upload.wikimedia.org/wikipedia/commons/f/f9/Bragg_Curve_for_ Alphas_in_Air-PT-en.svg, Creative Commons License.

Figure 1.8. *Rayleigh scattering.*

Source:https://upload.wikimedia.org/wikipedia/commons/c/c8/Rayleigh_scattering.jpg, Creative Commons License.

In photoelectric effect, (Figure 1.9) a photon is absorbed, resulting in the release of an electron from an atom or molecule. The atom undergoes ionization. The likelihood of this mechanism escalates swiftly with the atomic number of the interacting medium, generally proportional to Z^4. It plays a crucial function for photons with energies up to several 100 keV.

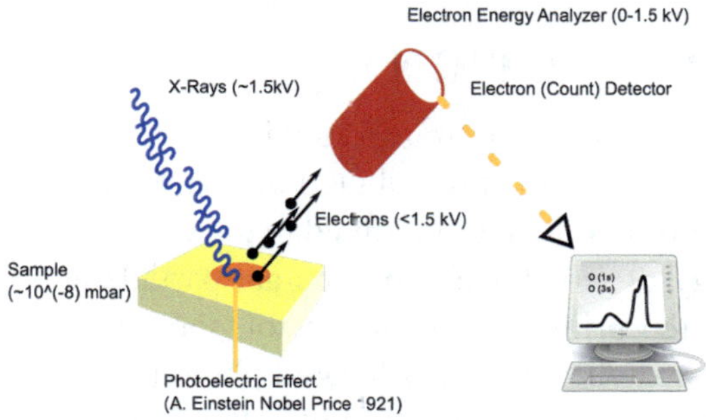

Figure 1.9. *Photoelectric effect.*

Source: https://upload.wikimedia.org/wikipedia/commons/8/83/XPS_CC_Roland_Siegbert.svg, Creative Commons License.

Compton scattering (Figure 1.10) involves the elastic scattering of a photon by a free electron. It is incoherent scattering, in which one electron departs from the atom, resulting in ionization.

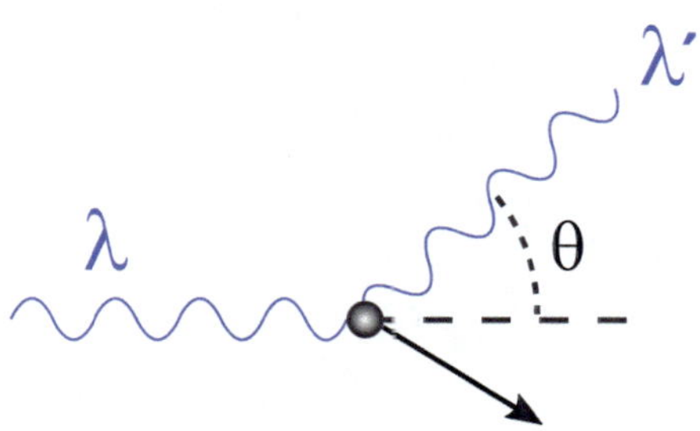

Figure 1.10. *Compton scattering.*

Source: https://upload.wikimedia.org/wikipedia/commons/e/e3/Compton-scattering.svg, Creative Commons License.

A photon remains post-interaction, however with less energy. Successive Compton collisions diminish photon energy, leading to the dominance of the photoelectric effect and, ultimately, the loss of photons. The Compton process approximately correlates with the atomic number (Z) of the medium. Compton scattering refers to the interaction in which a photon collides with a free electron within a material (Schumacher, 2005). The correlation of Compton scattering to atomic number Z is clear, as a material with atomic number Z has Z electrons. This process dominates for photons ranging from several 100 keV to approximately a few MeV in energy.

1.6. PAIR PRODUCTION

In pair production, a photon is transformed into a pair of an electron and a positron. The photon energy is distributed across the particle-antiparticle pair and the recoil electron or nucleus involved in the interaction. A photon requires a minimum energy of 1.022 MeV to generate particle-antiparticle pairs. This process becomes the primary mechanism for photon energy loss at energies exceeding approximately 10 MeV. Elevated photon energies and the presence of dense materials result in an increased likelihood of photons dissipating their energy by this mechanism, leading to their extinction. Pair production may involve two components. One arises from the interactions of photons with atomic electrons, whereas the other results from interactions with atomic nuclei.

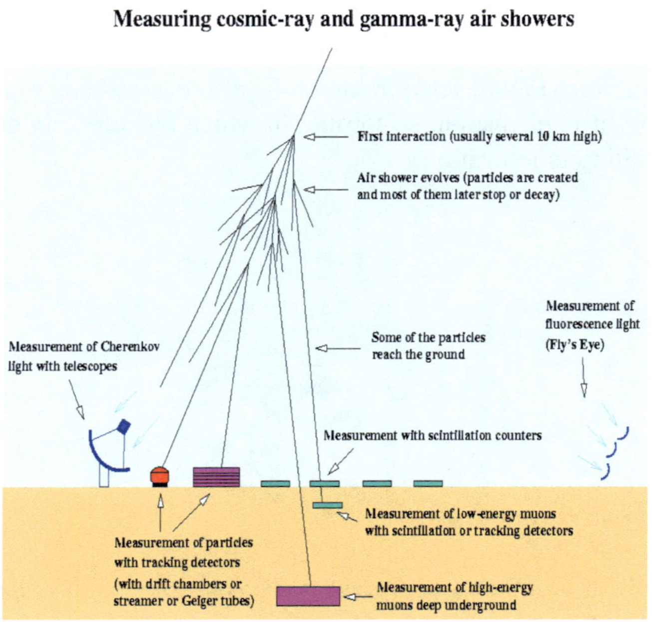

Figure 1.11. *Cherenkov light.*

Source: Bernlöhr, K. (1996). Measuring the UHE cosmic-ray composition with tracking detectors in air-shower arrays. Astroparticle Physics, 5(2), 139–146. https://doi.org/10.1016/0927-6505(96)00013-8, Creative Commons License.

The nuclear process dominates in high Z materials. Photo nuclear reactions refer to the interaction between photons and atomic nuclei. The likelihood of energy loss due to elastic or inelastic scattering from nuclei is minimal. At energy exceeding several MeV, photons can induce the dissociation of atomic nuclei, resulting in the emission of nuclear particles. The main type among them is neutron emission. This technique is crucial for all radiation safety and the design of nuclear facilities. The total intensity loss results from the cumulative effects of competing processes, which are determined by the probabilities (cross sections) of individual photons interacting through the mechanisms, including molecular, atomic, and nuclear interactions. The attenuation coefficient of a medium is the cumulative effect of all physical events that lead to the decrease in photon intensity. The assessment of mass attenuation coefficients gives data to evaluate diverse physical models of photon interactions with matter. Models compute cross sections of particular processes based on nuclear and quantum mechanical theories, allowing for the derivation of partial and total attenuation coefficients for specific materials throughout a finite range of photon energy. High-energy electrons in high-Z materials are more susceptible to energy loss by radiation. An electron is more prone to dissipate its energy by generating low-energy photons, indicating that each electron serves as a source of many photons. At each stage, an electron is diverted from its trajectory. High atomic number materials facilitate radiation more readily than their lighter counterparts. While we cannot determine the precise moment a photon interacts within a material, we can estimate the average energy loss of a photon beam. Cherenkov light (Figure 1.11) detection is widely employed as a method to differentiate and identify various types of charged particles (Forty and Ullaland, 2020). Cherenkov radiation is emitted in a homogeneous medium, whereas transition radiation occurs when a particle crosses the boundary between two mediums with differing dielectric constants. A dielectric medium is an ineffective conductor of electricity. The term dielectric is sometimes identical with insulator. The velocity of light in a medium is determined by the permeability (μ), which is based on the medium's magnetic properties, and the permittivity (μ), which pertains to its electrical qualities. For non-magnetic materials, the dielectric constant (k) directly quantifies the speed of light in a medium, and the relationship between the dielectric constant and the refractive index (n) is expressed as $k = n^2$. A medium puts force on radiation and charged particles. Charged particles release radiation when subjected to an external force. Synchrotron radiation denotes the radiation produced by electrons traversing a magnetic field (Lea, 1978). Although this name was originally coined for the light emitted by contemporary electron synchrotrons, a category of particle accelerators developed in the 20th century, the radiation from the Crab Nebula is also referred to as synchrotron radiation. The particles must be repeatedly bent to retrace their trajectory for more acceleration. A basic configuration comprises an electromagnet structured as a dipole, functioning as north and south poles, akin to the permanent magnets. The systems are advanced and complicated magnetic structures that guarantee the particles maintain their trajectory for many revolutions. A relativistic particle deflected in this manner generates radiation and dissipates energy, contradicting the fundamental objective of magnetic field configurations. The utilization of radiation as a resource for various purposes in research, technology, and industry soon became evident. Each charged particle possesses an accompanying radiation field. The energies of the fields fluctuate in accordance with the energies of the particles. While a particle traverses a linear path, the radiation and charged particle propagate simultaneously. An external field can deflect a charged particle without affecting its emission. The mathematics of synchrotron radiation emission is complex. The radiation loss, as well as the energy and intensity spectrum of the emitted radiation, is based on the particle velocity and the applied magnetic field B. We must subsequently modify these two parameters to attain the desired synchrotron light spectrum for the specific application. These magnetic configurations are referred to as undulators and wigglers. The phrases denote the degree to which electrons diverge from their designated trajectory. Undulators maintain minimal variances, generating high-intensity, virtually monochromatic beams. Wigglers generate high-intensity, high-energy photon beams that exhibit reduced monochromaticity. The radiation emitted by a bending magnet is continuous, with a high-frequency cutoff. A wiggler amplifies the intensity of high-energy photons by one to two orders of magnitude and truncates the low-energy spectrum. An undulator generates peaks of elevated intensities at particular energies.

1.7. NEUTRON

Neutron interactions with materials have characteristics akin to those of charged particles and photons, however certain elements of these interactions are unique to neutrons. Their interactions are the same as those of charged particles, as the neutron post-scattering remains the same as the initial one, but with reduced energy and a deviation from its original trajectory. In contrast to charged particles that create ionization trails along their trajectories, neutrons do not leave any trace. Similarly to photon propagation, they can traverse extensive distances without interactions and be absorbed at a single contact. Neutrons predominantly engage with atomic nuclei throughout a broad spectrum of energies. This behavior contrasts with the interaction of charged particles and photons, where energy loss or absorption primarily results from interactions with electrons. They also interact with atomic and molecule electrons at low energies, but to a lower extent. Due to their electrical neutrality, substantial mass, and small size, neutrons can infiltrate the cores of atoms even at low energies, approaching atomic nuclei closely and being quickly absorbed by them. The absorption is an exothermic process. It is typically succeeded by the release of photons, occasionally charged particles, and in certain instances, by the nuclear fission. The interactions generate radioactive nuclei that decay according to their characteristic half-lives and emit radiation. The energy loss of charged particles or photons during interactions exhibits a monotonic relationship with the atomic number (Z) of the medium. By understanding the atomic number and physical parameters such as density, we can determine which medium is more appropriate for an application. We can also delineate the quantity of energy deposits along the trajectory of charged particles and ranges. Aside from energy loss due to elastic scattering, neutron absorption is particular to the medium, influenced by both atomic number and mass number. The fractional energy loss due to elastic scattering changes consistently with the mass number of the medium, whereas neutron absorption is specific to the nuclear isotope contained inside the medium. Neutron interactions with materials significantly influence the selection of media for neutron moderation in nuclear reactors and the design of neutron detectors. We can use materials such as cadmium and gadolinium for neutron absorption, and ^{3}He or ^{10}B for neutron detection (Dumazert et al., 2018). The varieties of radiation, their intensities, and residual radioactivity levels are highly dependent on the interacting nucleus and neutron energy. This occurs because each nuclear isotope possesses distinct qualities, including excitation levels, quantum characteristics, and binding energy in relation to adjacent nuclei, which dictate the relative stabilities of nuclear systems. The cross section begins at approximately 100 barns for neutrons with energy of 1 μeV, drops to around 1 barn for 1 meV neutrons, and remains constant up to approximately 10 keV neutrons (Langel, 2023). Between 10 keV and about 1 MeV, the cross section exhibits significant variations, characterized by several maxima and minima. The maximum represents resonance processes in which a neutron is selectively absorbed to create a compound nucleus. It may be re-emitted or dissipated from the medium, leading to the emission of photons or other charged particles. The resonance locations and interaction strength are characteristics unique to the compound nuclear system. A compound nucleus formation is a reaction wherein a projectile is assimilated by the target nucleus to create a composite system. The composite system reaches a final state that exclusively depends on the characteristics of the composite compound nucleus. The resonance neutrons precisely describe the type of physical interactions that neutrons with energies ranging from a few eV to a few keV are likely to experience with materials. The resonances correspond to amplified interactions, frequently by a factor of 1,000 or greater compared to interactions at adjacent energy levels. They efficiently eliminate neutrons from the medium, while some resonant elastic scattering occurs as well. The term continuum neutrons pertain to the physics of neutron interactions. Neutrons are absorbed here as well, but the cross sections are far lower than those in the resonance zone. If resonances exist, they are exceedingly broad and extend across many MeV. The nuclear spectrum appears as a broad continuum. Neutrons have elastic collisions at all energy levels, which vary consistently with energy. If a neutron persists in a medium without undergoing absorption by nuclear processes, it will ultimately undergo beta decay, producing a proton, electron, and antineutrino. The half-life for this degradation process is 10.2 minutes. The neutrons are dissipated inside the medium through several mechanisms. We can examine several physical phenomena pertinent to each energy zone. At these energy levels, they may still be absorbed by atomic nuclei. The other phenomenon is scattering due to their contact with atomic electrons. This phenomenon is extensively utilized by scientists to investigate characteristics and conduct neutron imaging. A neutron incident on a ^{235}U target, if absorbed, creates a compound nuclear system of ^{236}U in an excited state (Bhatia et al., 2015).

Figure 1.12. *Nuclear reactor (interior).*

Source: https://upload.wikimedia.org/wikipedia/commons/6/6d/YAEDU_nuclear_reactor.jpg, Creative Commons License.

It leads to further distortion and initiates motion in the substance. The motion develops to the point that the neck fractures, yielding two nuclei and several neutrons. The nuclei generated at the fission site are unstable nuclei. They experience more beta decays until they attain pairs of stable nuclei. In a standard neutron-induced fission of a ^{235}U nucleus, two free neutrons are present. Free neutrons can initiate secondary fission events in the second phase, releasing four neutrons to induce fission in the third phase, and continuing this process. As the reaction induces other reactions, it is referred to as a chain reaction. If we eliminate neutrons, sustained reactions may not occur, resulting in the cessation of the process. A primary objective of nuclear reactor design is to regulate the quantity of neutrons and fissions to prevent disaster, while maintaining a stable fission environment. A reactor is deemed critical when each fission event results in one neutron initiating a subsequent fission. A reactor is deemed supercritical when more than one neutron induced fission, whereas it is classified as subcritical when less than one neutron is involved. When a thermal neutron is introduced, the resultant neutrons are typically not thermal. They are highly energetic neutrons. The fission cross-section for fast neutrons in ^{235}U is relatively small.

1.8. NUCLEAR REACTOR

A nuclear reactor, (Figure 1.12) whether a power reactor or a research reactor, is engineered to function in a regulated manner. This necessitates the design of the system to ensure that neutron interactions at every location within the reactor are identifiable and manageable. The objective is to arrange the materials within the reactor container to accurately estimate the quantity of neutrons and their energy distributions at any given time and location. The design needs considerable computer calculations; yet, the fundamental ideas are simple. We necessitate a sufficient quantity of slow (thermal) neutrons to guarantee continuous fission. Since fission neutrons are often rapid, it is necessary to decelerate them. During their progression towards lower energy, some would be absorbed. We must estimate this likelihood and confirm that sufficient neutrons are arriving at low energy to induce fission. This phenomenon is referred to as neutron transport within the medium. In nuclear reactors, the focus is mostly on the moderation and absorption of neutrons, as it is essential to create controllable chain reaction systems in a predictable manner. Moderators must effectively decelerate neutrons to thermal energy without absorbing them, whereas neutron-absorbing materials are necessary to halt the chain reaction at the operator discretion. The deceleration mechanism is achieved through the elastic scattering of neutrons with atoms in the medium. In elastic scatterings, the energy loss during a single interaction is based on the mass number of the medium. It can be deduced from Newtonian mechanics about elastic collisions. The capacity for moderation alone does not serve as the standard for assessing whether a medium is suitable for usage as a decelerating substance. If a medium possesses a substantial absorption cross section, it will function as a neutron poison, resulting in neutron loss through nuclear processes. The benefits of heavy water as a moderator are evident. Despite normal water possessing greater moderating capabilities, its absorption cross section, which exceeds that of heavy water by over a factor of 1,000, renders it a less favorable choice as a moderator. Although radiation risks are predominantly discussed in relation to living organisms, they also impact non-living substances, rendering transparent materials opaque, causing structural failure, and inducing many changes, including color changes. It can cause permanent or transient alterations in the compositions of these materials. A significant issue is the radiation damage to satellite instruments caused by cosmic radiation. The Van Allen radiation belts, spanning from 100 to 60,000 km above the Earth surface, are recognized areas of elevated energy electrons and protons, among other particles (Li and Hudson, 2019). In addressing radiation doses, it is essential to evaluate the cumulative impacts of energy deposits, ionizations, chemical changes, and biological effects. Physics can provide quantitative assessments of energy deposits and ionizations.

Multiple contributing factors exist. The effects of radiation are characterized by both stochastic and non-stochastic properties. Radiation damage varies according to the type of radiation, despite equivalent energy deposits and ionizations. Heavily charged particles inflict greater damage than photons or electrons for equivalent energy deposits. Various organs exhibit distinct responses to identical energy deposits. The effects of radiation are contingent upon the specific organs exposed to the doses. When a source emits multiple species or a single type of energy radiation, the effects are cumulative. It is necessary to utilize average energies and/or total energies of emitted radiations. When a radiation source is ingested, the resultant radiation damage is assessed by considering both the physical half-life of the radiation source and the biological half-life, which is the duration required for the body to eliminate the radiation. The international community has effectively established criteria for safeguarding from radiation. Exposure pertains to the capacity of incident radiation to ionize a material. For quantitative comparison, the ionization quantity of an air medium by radiation is utilized as a reference standard. The roentgen is the unit of exposure. The exposure level is defined as one roentgen if the radiation generates 2.58×10^{-4} coulombs of electric charge per kilogram of air (Pelt, 1999). In a radiation field, exposure can be estimated by utilizing an ionization chamber to measure the current flow within the chamber. One ampere of current traversing one kilogram or one milliampere of current through one gram of air corresponds to a 1 C/kg exposure rate. Cumulative exposure can be quantified by integrating charge across the duration of exposure. Dosage pertains to the energy assimilated by the medium. The SI unit is gray. The dosage is defined as one gray (Gy) when the energy deposition is 1 J/kg. Kinetic energy released per unit mass of the material (KERMA) is critically important in dosimetry. The energy emitted in the medium is responsible for the damage to the system. Collision loss refers to the localized deposition of energy into particles, mainly electrons, which subsequently dissipate energy. This phenomenon is referred to as collision KERMA (K_{col}). The second is radiative KERMA (K_{rad}) (Rogers and Townson, 2019). KERMA is applicable to photon beams and neutrons. Relativistic electrons exhibit photon-like behavior regarding energy loss, making KERMA applicable to them as well. It is necessary to assess the dose that a radiation source will provide to a target volume. It may involve a medical diagnosis utilizing X-rays or gamma rays, or it may pertain to radiation therapy. It may also be necessary to assess radiation levels to prevent unintentional exposure to elevated doses. The energy and intensities of gamma rays released by radioactive nuclei are characteristic of each radioactive isotope. Gamma ray attenuation fluctuates with their energy levels. For a specific radiation source, the relative contribution of each gamma ray is contingent upon the strength, referred to as branching per decay of the nucleus. Upon rectifying the radiation source, these parameters will be established. We must examine the attenuation of photons within the medium, which similarly fluctuates with energy levels. Upon defining the medium and the distance traversed by the radiation, the relative intensities and consequently the respective contributions of each photon energy to the dosage are also determined. For any radiation source and a certain medium, we can assign a specific gamma ray constant, sometimes referred to as the gamma factor. It was commonly accepted to utilize air as the medium, possibly due to the administration of ^{60}Co therapy or the implementation of X-ray diagnostics. Potential radiation sources include natural phenomena such as cosmic radiation, ambient background radiation, minor radioactive sources, particle accelerators, or nuclear reactors.

CHAPTER 2

NUCLEAR RADIATION

Contents

2.1. INTRODUCTION

The development of nuclear reactors for power generation in the United States commenced post-World War II. Engineers and scientists engaged in the atomic bomb construction recognized that the nuclear reactor would serve as a heat source for steam production, which could be utilized for power generation. Reactor projects commenced at Argonne National Laboratory (ANL) in Lemont, Illinois, and at Oak Ridge National Laboratory in Tennessee.

Figure 2.1. *Nuclear reactor.*

Source: https://upload.wikimedia.org/wikipedia/commons/c/cf/EBR-1%2C_August_2012.jpg, Creative Commons License.

The inaugural reactor project at NRTS was the Experimental Breeder Reactor-I (EBR-I) (Figure 2.1). The reactor construction commenced in 1949 and concluded in 1951. On December 20, 1951, a resistive load was linked to the reactor generator, producing around 45 kW of energy (Rachkov et al., 2014). The reactor could provide adequate electricity to meet the operational power requirements of the facility. The initial generation of power was accomplished via a sodium-cooled fast-breeder reactor. The Argonne National Laboratory (ANL) conducted the development of the boiling water reactor (BWR) (Figure 2.2).

Following the operation of multiple experimental reactors in Idaho, the Experimental Boiling Water Reactor (EBWR) was initiated in Illinois. The EBWR was the BWR power plant. The facility was originally operated at 5 Megawatts electric (MWe) and 20 Megawatts thermal (MWt). The commercial-scale Boiling Water Reactor was the Dresden Nuclear Power Plant. The pressurized water reactor (PWR) nuclear power plant constructed as a central station electrical producing facility was the Shipping-port Atomic Power

Station located near Pittsburgh, Pennsylvania (Solomon, 1984) (Figure 2.3). The facility generated 68 MWe and 231 MWt. It commenced operations in late 1957 and continued until 1982.

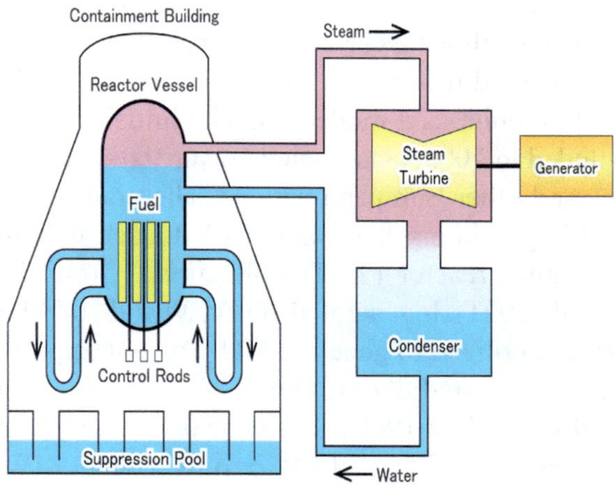

Figure 2.2. *Boiling water reactor.*

Source: https://upload.wikimedia.org/wikipedia/commons/4/4d/Boiling_Water_Reactor_01.svg, Creative Commons License

Figure 2.3. *Shipping-port atomic power station located near Pittsburgh.*

Source: https://cdn18.picryl.com/photo/2019/10/13/shippingport-atomic-power-station-on-ohio-river-25-miles-northwest-of-pittsburgh-36cc4c-1024.jpg, Creative Commons License.

Throughout its operational life, it functioned as a pressurized water reactor (PWR) and a light water breeder reactor (LWBR), with a core thorium blanket to produce ^{233}U as a reactor fuel. The Shipping-port reactor was derived from the reactor technology employed for naval propulsion. The initial gas-cooled reactor (GCR) power facilities were established in the United Kingdom. The initial models were cooled using CO_2. They utilized natural uranium metal as fuel and employed graphite as a moderator. The initial operation commenced in 1956 and concluded in 2003. Gas-cooled power stations were constructed in France, Germany, and other European nations. The uranium and thorium fuel were encapsulated in graphite spheres and cooled using helium. The thorium-fueled high-temperature reactor (THTR) functioned from 1985 to 1989 in Germany (Rabir et al., 2021). It generated 760 MWt and 307 MWe. The thorium in the fuel pellets was utilized to generate ^{233}U. Two GCR power stations have been functioning in the United States. The significant advancement of nuclear power commenced in the late 1960s. Power plants swiftly expanded in scale from a generating capacity of several MWe to over 1000 MWe. Construction and operation occurred globally. The advancement of nuclear energy was robust in the 1970s when the incident emerged at the Three Mile Island Unit 2 nuclear power facility outside Harrisburg, (Figure 2.4) Pennsylvania, in 1979 (Filburn and Bullard, 2016b). The reactor was a Pressurized Water Reactor.

Figure 2.4. *Three-mile Island unit 2 nuclear power facility.*

Source: https://upload.wikimedia.org/wikipedia/commons/2/27/Three_Mile_Island_Nuclear_Generating_Station_Unit_2.jpg, Creative Commons License.

2.2. ACCIDENTS

The accident paused reactor construction while the cause was investigated, leading to modifications in the design of reactors under construction to comply with new licensing standards. The repercussions of this incident were predominantly experienced in the United States. In 1986, an incident occurred at the Chernobyl Unit 4 reactor located near Kiev, Ukraine. The Chernobyl reactor (Figure 2.5) was a light water-cooled, graphite-moderated (LWG) reactor ((Stathis) Michaelides, 2012). This incident resulted in the emission of airborne radioactivity and the deaths of several responders.

Figure 2.5. *Chernobyl reactor.*

Source:https://upload.wikimedia.org/wikipedia/commons/8/87/IAEA_02790015_%2856131151 46%29.jpg, Creative Commons License.

Many nations with limited nuclear power initiatives discontinued their efforts in nuclear power generation. In 2010, the resurgence of the nuclear power industry appeared to be gaining momentum. Many governments were engaged in discussions and orders for new nuclear reactors. On March 11, 2011, a magnitude 9.0 earthquake occurred off the coast of Japan, subsequently followed by a 15-meter tsunami wave less than an hour later.

At the Fukushima Daiichi facility, (Figure 2.6) three of the six nuclear reactors, specifically units 1, 2, and 3, were operational, along with seven nuclear power facilities in the vicinity. All 11 operational nuclear power reactors in the region immediately ceased operations during the earthquake, and investigations revealed no substantial damage to any of the facilities as a result of the earthquake. The issue at Fukushima Daiichi resulted from the complete loss of both off-site and emergency power due to the tsunami, which damaged the on-site emergency generators necessary for maintaining cooling water circulation for four days. This incident affected the global nuclear power sector. Many nuclear power initiatives were suspended, and in certain states, it was decided to forgo the nuclear power option.

Figure 2.6. *Fukushima Daiichi facility.*

Source:https://upload.wikimedia.org/wikipedia/commons/4/4e/Fukushima_Daiichi_nuclear_plant._Two_types_of_above-ground_water_tanks_and_underground_pool.jpg, Creative Commons License.

By 2014, a total of 435 nuclear power reactors were operational globally. Many reactors may exist within a single power station, resulting in a lower quantity of power stations compared to reactors. The primary element of the reactor coolant system (RCS) is a robust reactor vessel that contains the nuclear core, its control rods, and essential components (Mehedinteanu, 2009). The container is cylindrical, including a hemispherical bottom (base) head. It is constructed from carbon steel, with all surfaces covered in stainless steel to mitigate corrosion. The internal core support and alignment structures are detachable to enable repair, as is the alignment structure for the top-mounted control rod mechanisms. The vessel inlet and outlet nozzles for the loops are positioned above the uppermost part of the fuel core. The nuclear core consists of many fuel assemblies organized into main areas to enhance fuel efficiency. All fuel assemblies are mechanically identical; however, the enrichment of the uranium dioxide fuel varies across the assemblies. In a standard initial core loading, three fuel enrichments are employed. Fuel assemblies with the highest enrichments are positioned in the core exterior, while groups of lower enrichment fuel assemblies are systematically arranged in the central region. During future refueling, 1/3rd of the fuel, representing the maximum burnup, is ejected, and new fuel is introduced into the outer portion of the core. The residual fuel is reorganized within the central 2/3rd of the core to attain optimal power distribution and fuel efficiency.

2.3. ROD CLUSTER CONTROL (RCC)

RCC assemblies utilized for reactor control contain absorber rods fastened to a spider connector, which is linked to a drive shaft (Schreiber, 2009). The control rods are infused with a substance possessing a high neutron absorption cross section. Above the core, control rods traverse within guide tubes that ensure the alignment of the control rods with thimbles (empty) of specific fuel assemblies at selected places in the core. RCC assemblies are regulated by a driving mechanism located on the reactor vessel head. The driving system enables the instantaneous release of the RCC assemblies (trip) when a rapid reactor shutdown is required. The assemblies are inserted during a trip via gravity. The objective is to normalize the power distribution across the core to the greatest extent feasible. Combined with control rods, there exists a configuration of absorber burnable poison (BP) rods attached to RCC-like structures, which are not linked to drive mechanisms. The BP rods stay in the core throughout operation but may be relocated after shutdown. Their objective is to mitigate the substantial surplus of nuclear reactivity during the initial phase of the cycle by utilizing the absorber throughout operation. They also permit a reduced concentration of soluble boron poison during operation. The configuration of each BP assembly resembles that of an RCC assembly, with the exception of the handling fitting. Positions in the cluster, where BP rods are absent, are filled by loose-fitting plugs that regulate the coolant flow throughout the host fuel assembly. The plugs are linked to the fixture. The fuel assemblies devoid of control rods, BPs, or neutron starting or startup sources contain pluggers. Pluggers are flow-balancing plugs attached to a fixture for support and manipulation. Specialized handling tools are required for each insert within a fuel assembly, as they are heated during operation but must be interchanged between assemblies. The elongated suspended rods are prevented from spreading by combs that maintain their correct alignment for reinsertion. All of these manipulations occur at significant depths underwater. Reactor coolant pumps (RCPs) are vertical, single-stage, mixed flow pumps characterized by a shaft-seal design. A flywheel on the pump motor shaft ensures extended coast-down durations to prevent abrupt reductions in core cooling flow during pump shutdowns. Interlocks and automatic reactor shutdowns guarantee the presence of forced cooling water flow anytime the reactor is operational. Two distinct power supplies are accessible to the pump motor when the plant is operational.

Steam generators include a vertical U-tube configuration with an enlarged upper part containing integrated moisture separation apparatus, ensuring steam quality of no less than 99.75% (Smith, 2022). Preheated feedwater enters the upper section of the unit, combines with effluent from the moisture separators, and thereafter descends along the exterior of the tube bundle. The feed is disseminated throughout the bundle and thereafter ascends alongside the heated tubes. The water in the secondary loop is vaporized in the lower area of the steam generator, and superheated in the upper section, eliminating the necessity for moisture separators prior to directing the dry steam to the turbines. The reactor coolant piping, and all pressure-containing and heat transfer surfaces in contact with reactor water are composed of stainless steel or stainless-steel cladding, with the exception of the steam generator tubes and fuel tubes, which are made of Inconel and Zircaloy, respectively.

2.4. REACTOR COOLANT SYSTEM (RCS)

An electrically heated pressurizer linked to one of the reactor coolant hot legs sustains RCS pressure during standard operation, mitigates pressure fluctuations during plant load transients, and ensures system pressure remains within design parameters under anomalous situations. A transient that may reduce system pressure is mitigated by the flashing of water within the pressurizer, which is maintained at saturation temperature by automated heaters. An escalating pressure transient is mitigated by the injection of cooler water from the primary loop into the pressurizer steam area to partially collapse the steam bubble, or through the automatic activation of relief and safety valves. The normal and emergency operation of the RCS necessitates various support functions to sustain water inventory, purify and treat primary coolant, eliminate residual heat post-plant shutdown, deliver cooling water to pumps and motors, supply ventilation air, and furnish emergency core cooling water. The transient power behavior of a nuclear core is dictated by a condition referred to as reactivity. In a core functioning at a constant power level, several factors influencing reactivity are balanced, resulting in a net reactivity of zero. A positive net reactivity results in an increase in power level, whereas a negative reactivity leads to a reduction. The regulation of a Pressurized Water Reactor (PWR) relies on the equilibrium of reactivity achieved through mechanical and chemical neutron absorbers, alongside suitable considerations for physical events affecting reactivity. The primary natural processes affecting transitory operation are the temperature coefficients of the moderator and fuel, as well as the accumulation or reduction of specific fission products. Reactivity balancing may emerge due to natural occurrences or the functioning of the reactor control system utilizing the RCCs. A variation in the temperature of the moderator or fuel may introduce or diminish reactivity and lead to alteration in power level until the reactivity change is balanced. RCC assemblies are utilized to manage substantial load transients, including load-follow operations, as well as during startup and shutdown processes. The chemical shim system employs the soluble neutron absorber boron, in the form of boric acid, which is introduced into the reactor coolant during cold shutdown, partially extracted at startup, and modified in concentration throughout the core lifespan to counteract effects such as fuel depletion and the buildup of fission products that inhibit the nuclear chain reaction. The control system enables the plant to accommodate step load increases of 10% and ramp load of 5% per minute within the load range of 15%–100% of full power, constrained by xenon restrictions towards the termination of core life (Kallappa et al., 1997).

2.5. FUEL RODS

The nuclear and T-G islands are entirely monitored from a centralized control room. A square arrangement or array of fuel rods that are structurally interconnected forms a fuel assembly. Control rod guide thimbles substitute fuel rods at designated intervals within the array and are secured to the upper and lower nozzles of the assembly. Spring clip grid assemblies are secured to the thimbles along the vertical axis of the fuel assembly to support the fuel rods. The fuel rods are secured and supported, with the distance between rod centerlines preserved inside this structural framework. The lower nozzle of the fuel assembly regulates the distribution of coolant flow and functions as the primary structural component at the base. The upper nozzle serves as the structural component of the fuel assembly and creates a plenum space where the heated reactor coolant is combined and directed towards the flow perforations in the upper core plate. The spring clip grids offer support for the fuel rods in two orthogonal or perpendicular directions. Each rod is supported at six locations within each grid cell. Each spring applies a force on the fuel rod that restricts lateral vibrations while permitting minor thermal expansions. The fuel rods are not physically secured to the support points, allowing them to extend axially to accommodate thermal expansion and radiation-induced development of the highly rough crystal structure of the Zircaloy cladding. RCC assemblies are employed for reactor initiation or cessation, to accommodate load fluctuations, and to regulate minor transient variations in reactivity. The control components of an RCC assembly comprise cylindrical neutron-absorbing rods or control rods that are nearly the same size as fuel rods and are interconnected at the top by a spider-like frame to create rod clusters. The control rods, composed of stainless-steel tubes containing a hafnium absorber substance, extend the entire length of the core when fully inserted. RCCs facilitate operational reactivity control and can consistently shut down the reactor, even when the most reactive RCC is immobilized outside the core. Each RCC is connected to its drive shaft, which is operated by a magnetically actuated drive mechanism attached to the reactor vessel head. RCCs are organized into clusters and electrically linked, enabling the collective movement of the entire group. The reactivity of the core is altered by a group inside the core. Each control rod of a RCC ascends and descends within its designated tubular guide thimble, which is an empty cylinder containing a constricted diameter near the base, a stopper/plug at the bottom, and perforations in the narrow part to facilitate coolant ingress and egress. Positioned symmetrically within fuel assemblies, these thimbles substitute fuel rods in the fuel assembly lattice. All fuel assemblies are constructed identically, as each may occupy a control rod position or any other location within the core. When an assembly is not in the RCC configuration, thimbles are either obstructed or contain BP rods, which are absorbers consumed during the reactor cycle. The thimbles function as guides for the control rods and operate as shock absorbers to decelerate control rod movement during reactor trips. In their completely retracted or withdrawn position, control rods do not exit the upper end of the guide thimbles. This guarantees that the rods remain consistently oriented and minimizes reactor coolant bypass via the thimbles. The design of RCC enhances core performance by offering a reasonably efficient method of regulation.

2.6. CONTROL RODS

Withdrawal of control rods from the core does not result in substantial power peaking because of the minor water gaps created. To mitigate the undesired effects of a substantial positive coefficient, the control given by chemical shim is diminished by incorporating aluminum oxide – boron carbide burnable absorber material within the core (Evans et al., 2022). Neutron sources for reactor initiation must be employed to elevate the neutron multiplication rate to levels measurable by external flux detectors, while remaining below criticality. The new fuel arrangement and the first low core reactivity due to random fissions alone would not allow for a securely regulated restart. Neutron sources are categorized into two types: a primary source, utilized for initial reactor startup and early operation of the first core, and a secondary source, employed for subsequent reactor starts and activated during reactor operation. The principal source is typically a fissionable isotope of californium. Secondary sources comprise a combination of antimony and beryllium (Sb – Be). The Sb undergoes radioactivity, generating high-energy gamma particles that remove neutrons from the Be. The primary and secondary sources resemble a control rod. Both varieties of source rods are encased in stainless steel. The secondary source rods comprise Sb – Be pellets that are not initially active. The principal source rods encompass sealed capsules of source material located at axial locations. The cladding encapsulation is finalized by seal welding the end plugs. The precise core position of the sources is established during the final design phase to ensure sufficient neutron flux at the source range detectors consistently. Reactor coolant piping and fittings are constructed from stainless steel or carbon steel encased in stainless steel. Carbon steel is utilized for the pressurizer relief line, linking the pressurizer safety and relief valves to the flanged nozzle of the pressurizer relief tank, as well as for the nitrogen supply, vent, and drain lines associated with the pressurizer relief tank. The pressurizer surge and spray lines, loop drains, and interconnections to other systems are constructed from stainless steel. All joints and connections are welded, with the exception of the flanged pressurizer safety valve outlet nozzles. Thermal sleeves are inserted in locations susceptible to substantial thermal stresses due to fast fluctuations in fluid temperature during transients. Valves, pipelines, and equipment operating at elevated temperatures are typically insulated thermally to minimize heat loss. The insulation encasing the piping and components of the RCS is engineered to enable its removal for routine inspections. The insulation utilized for the RCS is explicitly mandated to restrict chlorides and other halogens. Reactor vessels are often insulated using reflective metal insulation systems. The feed rate to the RCS is regulated automatically by the pressurizer water level, however the bleed rate can be adjusted by selecting the appropriate mix of letdown orifices to satisfy plant operational needs (Fynan and Park, 2021). The water exits the RCS and traverses the shell side of the regenerative heat exchanger, where it transfers its heat to the makeup water returning to the RCS. The water thereafter traverses orifices, where its pressure is diminished, followed by passage via a nonregenerative heat exchanger, and settles with an additional pressure decrease via a low-pressure valve. Following its passage through a mixed-bed demineralizer that eliminates ionic contaminants, the water proceeds through the reactor coolant filter and enters the volume control tank (VCT) through a spray nozzle. A secondary route downstream of the mixed-bed demineralizers can be utilized to channel the flow to the boron recycle system (BRS). The vapor space in the VCT contains hydrogen, which dissolves in the coolant and establishes the hydrogen concentration in the RCS. Fission or other non-condensable gases (He, H_2) present in the system are eliminated through continuous, intermittent, or pre-shutdown of the VCT. The continuous purging of the VCT significantly decreases the activity level of the reactor coolant. The charge pumps extract coolant from the VCT or the BTRS and direct it along two parallel routes: returning to the RCS via the tube side of the regenerative heat exchanger and supplying the seals of the RCPs. The RCP seal injection flow enters the pump between the labyrinth seals and the seal. The flow bifurcates, with a portion entering the RCS and the remainder exiting the pumps as regulated seal leakage. The controlled leakage water from the pumps is directed to the seal water heat exchanger before returning to the suction of the charge pump for recirculation. Water introduced into the RCS through the RCP seals is redirected to the charging pump suction via the heat exchanger and the seal water heat exchanger (Hsu et al., 2020). Increases in the RCS volume is managed by the pressurizer. The VCT is engineered to accommodate for pressurizer level discrepancies. When the VCT reaches capacity, a high-water level indicator redirects flow to the BRS or waste processing system. Low-level signals from the VCT activate reactor makeup control or flow from the refueling water storage tank (RWST) as a safety measure. Makeup for the RCS may originate from the sources given here. (a). A supply of demineralized and deaerated water is used to diminish the concentration of dissolved boric acid in the chemical shim to adjust for core burnup; (b). Chemical mixing tank for the addition of minor volumes of hydrazine for oxygen scavenging or lithium hydroxide for pH regulation; (c). RWST for the emergency replenishment of borated water. d. Boric acid sourced from the boric acid tanks; and (d). A mixture of demineralized and deaerated water with concentrated boric acid to achieve the reactor coolant boron concentration for standard plant makeup. A continuous supply of clean water and a concentrated boric acid solution must be supplied to replace these resources. A secondary flow route downstream of the VCT channels the suction flow of the charging pump to the BTRS to facilitate alterations in boron concentration during load-following and other operational activities of the plant. A boron concentration meter offers precise measurement of boron levels in the reactor coolant.

2.7. STANDARD OPERATIONS

The readout offers control data that the plant operator can employ in standard operations involving makeup, let down, chemical treatment of reactor coolant, and the adjustment of boron concentration. This data is utilized in a boron follow program to aid the operator during fluctuations in boron concentration, such as load follow, hot standby, fuel burnup, and cold shutdown. Sample sites or points for the measurement unit are situated both upstream and downstream of the boron thermal regeneration demineralizers. Concentrated boric acid is processed through a demineralizer to eliminate water-soluble contaminants, including aluminum, calcium, and magnesium. A filter is supplied for the elimination of suspended particles. Dilution is attained by reducing the concentration of boric acid in the reactor coolant by the transfer of flow from the VCT, at comparatively low temperatures (10°C), to the thermal regeneration demineralizers (Manno and Golay, 1985). The resin, previously depleted of boron at elevated temperatures during a boration process, is now able to absorb boric acid from the low-temperature effluent of the VCT. The reactor coolant, having a reduced boron concentration, exits the demineralizers and is routed to the CVCS. During operation, the charging pump suction stream enters the tube side of the moderating heat exchanger, bypasses the letdown cold heat exchanger, and traverses the shell side of the letdown reheat heat exchanger. The moderating and letdown reheat heat exchangers elevate the BTRS stream to 140°F prior to its entry into the demineralizers. The temperature of the BTRS stream at the demineralizer entry point is regulated automatically by a temperature control valve that manages the flow rate on the tube side of the letdown reheat heat exchangers. The BTRS stream traverses the demineralizers, enters the shell-side of the moderating heat exchanger, flows through the tube side of the letdown chiller heat exchanger, and proceeds to the charging pump suction. The temperature of the BTRS stream entering the charging pump is regulated automatically by the temperature control valve, which manages the shell-side flow rate in the letdown chiller heat exchangers. For boration, the concentration of boric acid in the reactor coolant is elevated by directing the flow from the VCT, at elevated temperatures, to the thermal regeneration demineralizers. The water passing through the demineralizers discharges boron that was retained on the resin at low temperatures during a prior dilution process. The boron-enriched reactor coolant is returned to the Reactor Coolant System through the Chemical and Volume Control System (CVCS). The BTRS recirculation pump generates the necessary force to direct flow from the VCT to the demineralizers during boration and dilution phases. When the BTRS is not utilized for boration or dilution, it is separated from the CVCS, allowing the recirculation pump to function and adjust the temperature of the demineralizer beds for the operational mode.

2.8. HEAT REMOVAL SYSTEMS

The principal role of the residual heat removal system (RHRS) is to carry thermal energy from the core and RCS during plant cooldown and refueling activities (Zou et al., 2015). The technology is engineered to lower the RCS temperature to 140°F within 20 hours post-reactor shutdown. The RHRS can also facilitate the transfer of refueling water between the refueling cavity and the RWST during the initiation and end of refueling operations. The residual heat removal pumps and heat exchangers are employed as components of the safety injection system for emergency core cooling during a loss-of-coolant accident (LOCA). The configuration of two-loop and three-loop plants varies solely in the dimensions of the components and the quantity of branch injection lines to the RCS. The RHRS comprises two independent, redundant mechanical subsystems, each receiving electrical power from one of two distinct and redundant electrical power trains. Each subsystem comprises one RHR pump, one RHR heat exchanger, and the necessary pipes, valves, and instrumentation. The RHR pumps and heat exchangers are situated in the auxiliary building, positioned as near as feasible to the containment. The plumbing configurations of the two subsystems are similar, with no significant piping connections between the subsystems in the auxiliary building. Each subsystem features a suction line from an RCS hot leg, equipped with two closed motor-operated series isolation valves that ensure compliance with RCS pressure boundary and containment isolation standards. Each subsystem draws from the containment sump and the Refueling Water Storage Tank (RWST). The RHRS necessitates these suction sources for its functionality during the mitigation of a LOCA as part of the SIS (Kim et al., 2015). The discharge line of each pump is directed to its corresponding RHR heat exchanger and then into the containment. A bypass loop is employed around the RHR heat exchanger to facilitate the regulation of RCS cooling while ensuring a consistent overall pump flow. A minimum flow for the RHR pump is ensured by a recirculation line from the heat exchanger outlet to the pump suction, which remains open in all operational modes of the system. Within the containment, each pump discharge is split into four branch lines. Each branch line supplies its corresponding cold leg via the designated accumulator discharge line. No cross connects exist between the two subsystems before the discharge headers inject into the accumulator discharge lines. The branch lines within the containment are separated or isolated from the RCS cold legs by check valves. During system operation, each RHR pump draws suction from one of the RCS hot legs via its own suction line. The pumps subsequently expel flow through the RHR heat exchangers, which facilitate the transfer of heat from the heated reactor coolant fluid to the component cooling water running through the shell side of the RHR heat exchanger. The cooled RCS flow is returned to the RCS cold legs via the shared SIS piping. The steam generator blowdown processing system (SGBPS) treats blowdown from steam generators to comply with chemical specifications for recycling into the main condenser and radiochemical specifications for environmental discharge when necessary. During standard operation, blowdown from each steam generator is sent into a heat exchanger, where the temperature is diminished by condensate water. The pressure is diminished, and the blowdown is channeled through the prefilter and mixed-bed demineralizers sequentially. The fluid traverses a radiation monitor and is often recirculated to the main condenser, but may be released into the environment via the discharge canal when necessary. Instrumentation is installed at key locations to assess the functional integrity and operational efficiency of the system. An elevation in conductivity, observed at the exit of each demineralizer, signifies resin bed depletion. When the upstream bed is depleted, flow is redirected around that bed. The spent resin sluice pump facilitates the transfer of spent resin to the storage tank, while new resin is introduced into the demineralizer. The newly regenerated demineralizer is subsequently reintroduced as the new downstream bed. Prior to resin disposal, the used resin storage tank is fluidized with nitrogen gas or sluice water to remove the resin. The tank is pressurized with nitrogen to expel the resin into the solid waste processing system.

Figure 2.7. *Power plant.*

Source: https://upload.wikimedia.org/wikipedia/commons/4/4e/Nuclear_Power_Plant_Cattenom.jpg, Creative Commons License.

2.9. POWER PLANT

The power plant (Figure 2.7) incorporates several barriers that ensure the containment of radioactive materials at essential levels like Zircaloy tubes of the fuel rod assemblies, RCS pressure limits (boundary), and Reactor containment structure (building). The structure offers biological protection for both routine and emergency scenarios. Various types of containment structures have been developed. The commonly utilized designs feature steel vaults or concrete containers lined with steel plating. Steel vessels may possess either a cylindrical or spherical configuration. Reinforced concrete vessels, which may occasionally be post-tensioned, are cylindrical with hemispherical domes. The interior surface of the concrete shell, dome, and floor is coated with steel plates to guarantee a high level of leak tightness. The cylindrical shell is subjected to prestress by a post-tensioning system including horizontal and vertical tendons. The dome features a bidirectional post-tensioning system. The foundation base slab is a concrete construction typically reinforced with high-strength steel. A continuous access gallery is available beneath the base slab for the installation and inspection of vertical tendons. The containment fully encases the reactor and RCS, guaranteeing that the permissible threshold for radioactive material leakage into the environment will not be surpassed, even in the event of a catastrophic RCS failure. The enclosure is engineered to withstand all probable loading circumstances, including standard loads, loads during Loss of Coolant Accidents (LOCA), test loads, and loads resulting from adverse environmental factors. A crucial element of the safety-related systems is the hydrogen detection and ignition system. The interaction of steam with Zircaloy-4 fuel cladding may generate hydrogen within the containment in the event of a pipe rupture or other venting of RCS coolant beyond the pressure boundary (Zuidersma et al., 2024). An explosive combination may occur if the hydrogen content in dry air attains around 20%. The threshold increases when a substantial quantity of steam is also present. The purpose of the hydrogen control system is to identify hydrogen concentrations below 20% and mitigate it by reacting with air in catalytic heaters positioned at some locations within the containment, ensuring sufficient mixing and averts the formation of an explosive mixture. Zircaloy-4 is an alloy composed of zirconium, around 1.5% tin, and trace quantities of iron, nickel, chromium, and around 200 ppm of carbon. It serves as a window for neutrons, develops an oxide layer to inhibit corrosion, and its mechanical characteristics ensure sufficient strength and ductility. It does not engage with uranium dioxide fuel and has minimal reactivity with fission products. Substantial ingots of the alloy are divided into trexes of a few inches in thickness and several inches in diameter (Johnson et al., 2024). A hole approximately one inch in diameter is centered in the trex. A mandrel exerts pressure from the rear, refining the interior surface as the material elongates, and reduces in diameter through multiple stages until it transforms into tubing. The Pilger technique involves a pair of mandrels that compress the material at a single location, after which the tube is slightly rotated, and the procedure is repeated. Zirconium quickly undergoes cold working and requires annealing between stages to preserve ductility for subsequent processing operations. The 1979 Three Mile Island incident was intensified by ambiguity regarding the presence of a steam bubble in both the reactor vessel head and the pressurizer. The crew hesitated to pressurize the coolant, fearing it could lead to system rupture. The time delay caused partial core melting due to insufficient emergency cooling. Following that incident, it has become clear to remotely vent the reactor vessel head to ensure the absence of any bubbles during an accident or at any time. The specifications for high point vents are delineated by the reactor vessel head vent valves and the ADS valves. The ADS system has been delineated. The principal purpose of the reactor vessel head vent is to facilitate the correct filling of the RCS and vessel head during plant startup. The reactor vessel head vent valves and the ADS valves can be actuated and managed from the main control room. The reactor vessel head vent valves facilitate the expulsion of non-condensable gases or steam from the reactor vessel head to alleviate potential inadequate core cooling or compromised natural circulation through the steam generators caused by the buildup of non-condensable gases in the RCS. The reactor vessel head vent system design should comply with NRC regulations. The reactor vessel head vent valves can be actuated from the main control room to establish an emergency letdown pathway, utilized to avert pressurizer overfill during prolonged heat sink loss incidences. An orifice is installed downstream of each set of head vent valves to restrict the emergency letdown flow rate. The initial-stage valves of the ADS are connected to the pressurizer and enable the extraction of non-condensable gases from the pressurizer steam area post-accident. The release of non-condensable gases from the pressurizer steam area is unnecessary for ensuring safety-related core cooling after a hypothetical accident. Gas accumulations are eliminated through the remote manual control of the first-stage ADS valves. The PRHR heat exchanger piping and the core makeup tank intake piping in the PXS are equipped with high point vents that facilitate the removal of non-condensable gases, which may disrupt the operation of the heat exchanger or core makeup tank (Juhn et al., 2000). Non-condensable gases that accumulate in these elevated areas can be manually evacuated. The discharge from the PRHR heat exchanger high point vent is directed to the IRWST. The discharge from the core makeup tank high point vent is directed to the reactor coolant drain tank. The reactor vessel head vent configuration is intended to eliminate non-condensable gases or steam from the RCS through remote manual operations conducted from the main control room via a pair of valves. The system releases to the IRWST. The reactor vessel head vent system is engineered to offer an emergency letdown pathway to avert prolonged pressurizer overfill after heat sink loss incidents. The reactor vessel head vent is engineered to restrict the emergency letdown flow rate to the operational capacity of the standard makeup system. The reactor vessel head vent system is capable of emitting non-condensable gases from the reactor head after a major accident. The system exhausts the reactor vessel head utilizing exclusively safety-related apparatus. In addition to passive systems within the containment, there exist passive systems for cooling the containment itself when it is filled with steam and exhibits high-pressure signals. All passive systems are designed to contain radioactivity and safeguard the fuel and associated equipment from additional deterioration in the long term after the Design Basis Accident (DBA). Access to the reactor by staff is presumed to be prohibited throughout this period of around 1.5 months. The apparatus serves other purposes under both normal and abnormal conditions; yet, its safety function is uncompromised by these alternative applications, like fire prevention or refueling water makeup.

The passive containment cooling water storage tank (PCCWST) is integrated into the building structure above the containment vessel. The interior surfaces of the tank are clad with stainless steel plates. It contains demineralized water and possesses the minimum requisite usable volume for the passive containment cooling function, exceeding 756,000 gallons. The passive containment cooling system serves as the heat sink. The passive containment cooling water storage tank is engineered for seismic resilience and missile protection. The adjacent reinforced concrete supporting structure contributes to the safeguarding of the passive cooling system.

2.10. MANAGEMENT

A recirculation loop for chemistry and temperature management is linked to the tank to ensure system operability. The tank is designed to ensure adequate thermal inertia and insulation, allowing for draindown to occur without the need for heater operation. The PCCWST suction pipe for the fire protection system is designed to ensure that activation of the fire protection system does not compromise the usable capacity designated for the passive containment cooling function. The system actuation involves the opening of the isolation valves for the passive containment cooling water storage tank. This enables the passive containment cooling water storage tank to supply water to the exterior surface of the steel containment shell. The gravitational pull facilitates the flow of water, creating a film across the dome and side walls of the containment structure. The cooling mechanism reduces the pressure within the confinement. The boiling water reactors (BWRs) nuclear plant, similar to the pressurized water reactor (PWR), originates from the technology developed in the 1950s for the United States Navy nuclear submarine program (Theriault, 2016). The primary distinction between the PWR and BWR is that the latter operates as a direct cycle nuclear plant, wherein heat is generated in the fuel region and water vaporizes inside the walls of the fuel bundles. The BWR design has been streamlined in two primary aspects: reactor systems and containment architecture. Steam was produced in the reactor, then transferred to an elevated steam drum and a secondary steam generator, before proceeding to the turbine. General Electric identified the BWR as the advantageous nuclear power concept because of its intrinsic benefits in control and design simplicity, subsequently establishing an atomic power equipment division in 1955 to market it commercially. Except for its heat source, the BWR generation cycle is largely analogous to that of fossil-fueled power plants. Fine-grained steels and sophisticated fabrication methods are chosen to optimize the structural integrity of the vessel. BWR vessels exhibit the minimal neutron exposure among all light water reactors, and the annular space that transports recirculating water and feedwater downward between the core shield and the vessel mitigates radiation exposure to the vessel wall material. Samples of vessel material are positioned throughout the vessel to facilitate regular monitoring of exposure and material characteristics. The vessel head closure seal has two concentric metallic O-rings. This sealing system has been shown to operate without any detectable leakage under all conditions. A leak detection system is employed to monitor seal integrity. The vessel supports and internal supports are engineered to withstand combined loads, encompassing control rod drive reactions, and jet reaction thrusts. The vessel is attached to a supporting skirt, which is fastened to a concrete and steel cylindrical joint with the reactor building foundation. The design of the vessel and its accompanying piping includes many features to facilitate the refueling process. Steam outlet lines are welded to the vessel body, thus avoiding the necessity to dismantle flanged connections in the steam lines during head removal for refueling. Another design element is the seal between the vessel and the adjacent drywell.

2.11. CORE

The shroud is a cylindrical stainless-steel entity that encases the core, serving as a barrier to distinguish the upward flow through the core from the downward flow to the annulus. The jet pump discharge diffusers extend through the peripheral shelf of the shroud support beneath the core elevation to deliver coolant into the inlet plenum. The peripheral shelf of the shroud support is welded to the vessel wall to prevent the jet pump output flow from circumventing the core and to create a chamber surrounding the core, which can be re-flooded during a loss-of-coolant accident (LOCA). The shroud support bears the weight of the shroud, steam separators, jet pump system, and seismic and pressure stresses during both normal and fault operational conditions. Two ring spargers, designated for LPCS and HPCS respectively, are attached within the core shroud in the area between the upper section of the core and the base of the steam separator (Theriault, 2009). The core spray ring spargers are equipped with spray nozzles for the injection of cooling water. The core spray spargers and nozzles do not obstruct the installation or extraction of fuel from the core. A nozzle for injecting the neutron absorber solution (sodium pentaborate) is positioned beneath the core in the area of the recirculation inlet plenum. The steam separator assembly has a domed base upon which an array of standpipes is welded, with a three-stage steam separator positioned at the apex of each standpipe. The steam separator assembly is positioned on the upper flange of the core shroud and constitutes the cover of the core discharge plenum area. The seal between the separator assembly and core shroud flanges is a metal-to-metal contact and does not necessitate a gasket or alternative sealing mechanisms. The separator assembly is attached to the core shroud flange by a hold-down bolt, facilitating removal and extension above the separators. During installation, the separator base is aligned with the core shroud flange using guide rods and secured in place with locating pins. The long-bolt design aims to facilitate direct access to the bolts during reactor refueling operations, minimizing underwater tool manipulation for the removal and installation of assemblies, hence eliminating the need to engage threads while securing the shroud head. The final loading is achieved via the differential expansion of the bolt and the compression sleeve. The fixed axial flow steam separators include no moving components and are constructed from stainless steel. The steam subsequently exits the separator at the apex and enters the wet steam plenum beneath the dryer. The steam dryer assembly is attached within the reactor vessel, positioned above the steam separator assembly, and constitutes the upper and lateral boundaries of the wet steam plenum. Vertical guides within the vessel facilitate the alignment of the dryer assembly during installation. Inward-extending pads from the vessel wall support the dryer assembly, which is secured in situ during operation by the vessel head. Steam from the separators ascends and disperses via the drying vanes. The vanes are attached to upper and lower supporting components, creating a cohesive structural unit. Moisture is extracted and transported by a network of drains to the pool around the separators, thereafter flowing into the recirculation downcomer annulus.

CHAPTER 3

RADIATION PHYSICS AND HEALTH

Contents

3.1. INTRODUCTION

The nuclear fuel cycle (Figure 3.1) has effectively enriched uranium for reactor fuel and weapons manufacturing, as well as recycled spent nuclear fuel to extract uranium and plutonium (Rodríguez-Penalonga and Moratilla, 2017). The enrichment process necessitated extensive facilities due to the inefficient diffusion and centrifuge methods for uranium enrichment. The emergence of sophisticated centrifuge technology and laser isotope separation significantly enhances the efficiency of uranium enrichment and enables the establishment and operation of smaller plants.

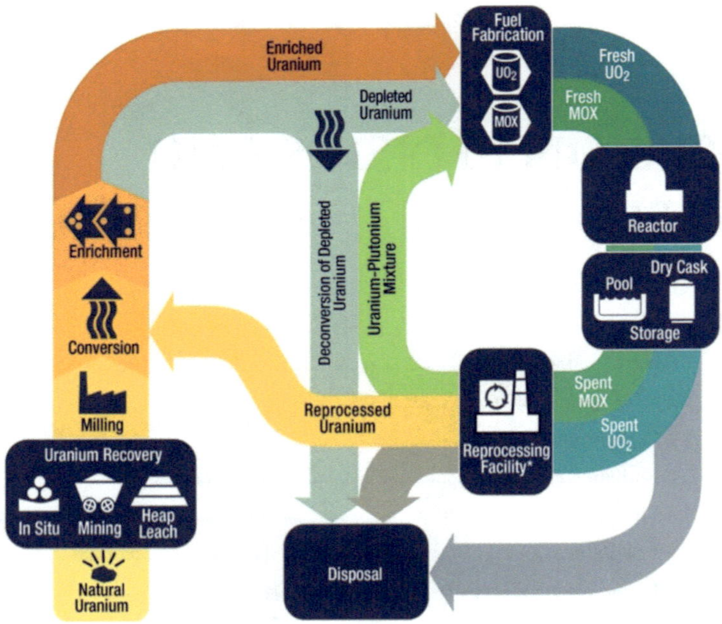

Figure 3.1. *Nuclear fuel cycle.*

Source: https://upload.wikimedia.org/wikipedia/commons/f/f1/The_Nuclear_Fuel_Cycle_%2844021369082%29_%28cropped%29.jpg, Creative Commons License.

These facilities are more easily concealed than the huge centrifuge and gaseous diffusion facilities. This creates the potential for a concealed enrichment facility to generate weapons-grade uranium. Advanced technologies, especially laser uranium enrichment, pose a nuclear proliferation risk in the 20th century. Reprocessing technology has effectively retrieved plutonium, and this technique is widely recognized. The proliferation of nuclear power plants presents the risk of spent fuel being reprocessed, with the extracted plutonium potentially redirected for other activities.

Nuclear medicine (Figure 3.2) has advanced and enhanced diagnostic and therapeutic approaches. The capability to localize the absorbed dose has improved, and different radiation types are being applied to target malignancies. Proton and heavy ion therapy approaches are prevalent, and initial investigations utilizing antiprotons have been attempted. Nanotechnology and internal radiation-generating devices are being developed for targeted dosage administration in cancer therapeutic applications. The progression of nuclear medicine technology has elevated the average absorbed dose administered to the population. The utilization of nuclear elements in commercial products and their unintentional incorporation into scrap metal utilized in consumer goods provide more concerns.

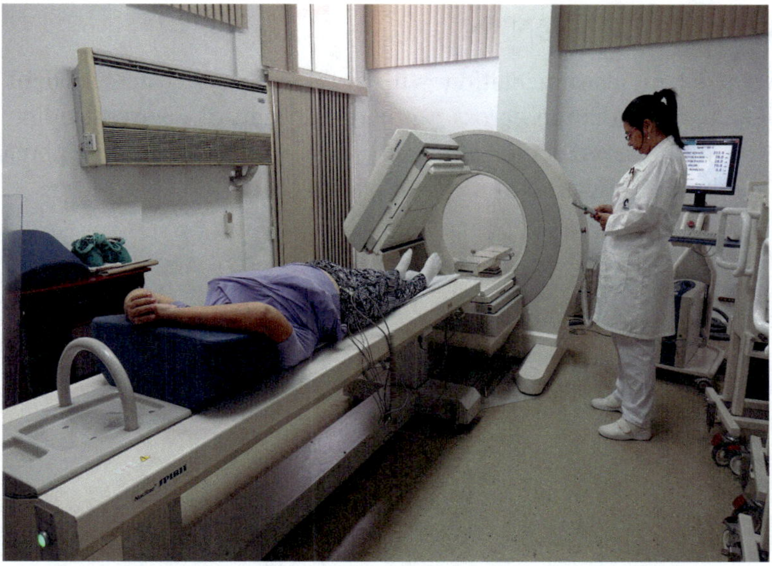

Figure 3.2. *Nuclear medicine.*

Source:https://upload.wikimedia.org/wikipedia/commons/e/e6/Nuclear_Medicine_in_Cuba_%2808010596%29_%2842623325854%29.jpg, Creative Commons License.

Public concerns about nuclear power generation and the repercussions of significant accidents have intensified due to the Fukushima Daiichi incident and its exaggerated portrayal by the media and antinuclear organizations. Incidents related to radioactive materials and the subsequent media coverage indicate that public interest in radiation-generating technology and radioactive substances is expected to rise. The media poses a considerable issue as its viewpoint is frequently swayed more by emotion and sensationalism than by scientific rationale and expertise. Significant regulatory measures have been implemented following the Fukushima Daiichi incident, impacting both operational plants and those in the design and building phases (Alvarenga and Frutuoso, 2015). The twentieth century witnessed the growth of health physics and its scientific foundation, whereas the 21st century will necessitate further scientific education for health physics experts to address the substantial difficulties presented by new technology. National and international organizations persist in promoting ongoing development and standardization; yet, they risk becoming disconnected from applied health physics due to controversies around the linear-nonthreshold (LNT) hypothesis and environmental protection.

3.2. ADVANCEMENTS

Advancements in instrumentation will enable improved detection of many types of ionizing radiation across a broad spectrum of energy, leading to the integration of these detectors into consumer products like cell phones and enhancing the identification of illegal nuclear materials.

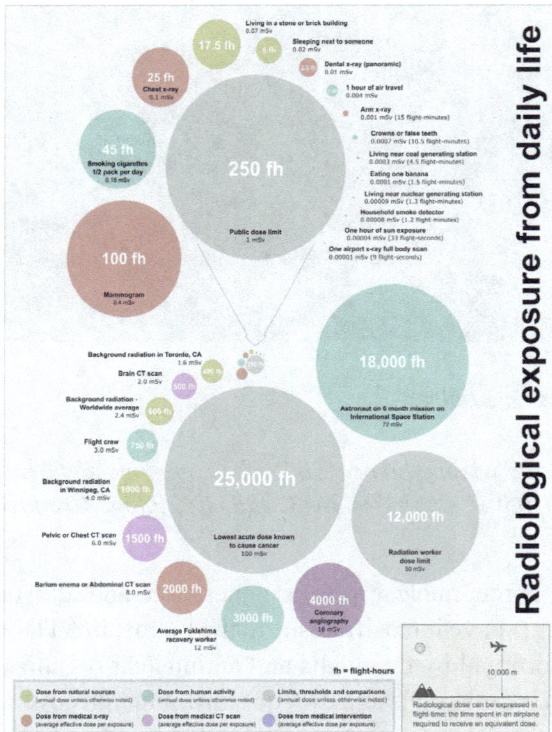

Figure 3.3. *Radiation exposure.*

Source: https://upload.wikimedia.org/wikipedia/commons/5/52/Radiological_exposure_from_daily_life.png, Creative Commons License.

Health physics is an essential discipline with a promising future. Considerable obstacles are expected to emerge when novel technologies increase the utilization of radioactive materials and radiation-emitting equipment, malfunctions of current technology emerge, radical activities extend to encompass radioactive materials or nuclear armaments, and outdated models disintegrate. Health physics is closely connected to the advancement of nuclear technology and nuclear-related occurrences. This connection will be evident in conventional domains and potentially in emerging sectors, such as public space tourism and nuclear terrorism. Engagement with stakeholders and the public is crucial to mitigate misunderstanding and panic that frequently accompany media coverage of nuclear-related incidents. The health physicist of the 21st century must possess technical proficiency and the ability to convey information to the public in a clear and comprehensible way suitable for an audience with minute scientific understanding. The National Council on Radiation Protection and Measurements postulates that improvised nuclear bombs will not surpass 10 kilotons (TNT equivalent) (Sanders, 2017). The 10 kT design assumption should be broadened to encompass larger weapon yields for the formulation of procedures, and resource allocations. The 10 kT limit apparently prevents the potential for the theft of an existent device from a nuclear power, the transfer of a device from a nuclear power to a terrorist organization, or the utilization of established scientific resources to create a higher-yield covert device.

3.3. RADIATION EXPOSURE

The majority of public radiation exposure originates from natural background sources and nuclear medical treatments (Figure 3.3). When recognized by the public, these radiation levels are deemed more acceptable than those obtained from alternative sources. Public attention is occasionally stimulated by emerging technologies. Nuclear medicine is primarily categorized into imaging or diagnostic methods and therapeutic regimens or protocols (Taïeb et al., 2019).

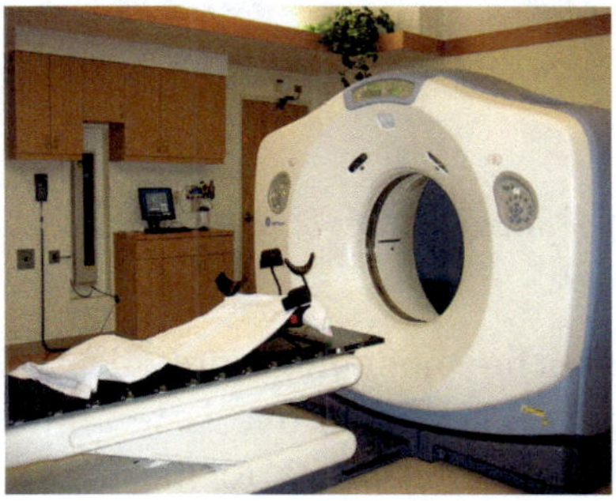

Figure 3.4. *Computed tomography.*

Source: https://upload.wikimedia.org/wikipedia/commons/d/d8/CAT_scanner.jpg, Creative Commons License.

Imaging includes computed tomography (CT), (Figure 3.4) radiography, fluoroscopy, tracer studies, and the injection of radiopharmaceuticals, all aimed at acquiring information without damaging healthy tissue. These imaging modalities complement other techniques such as optical imaging, ultrasound (US), and magnetic resonance imaging (MRI) (Figure 3.5) to collect data for subsequent assessment and therapy development or planning.

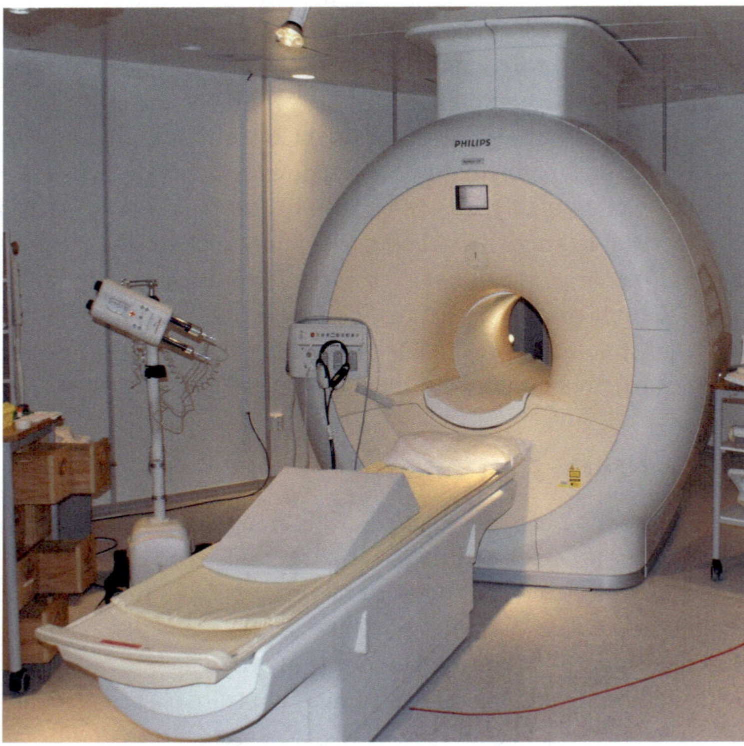

Figure 3.5. *MRI.*

Source: https://upload.wikimedia.org/wikipedia/commons/e/ee/MRI-Philips.JPG, Creative Commons License.

Nuclear medicine treatment modalities provide alternatives to invasive surgery and chemotherapy. Their objective is to eliminate infected tissue by preferentially delivering absorbed doses to the tumor location. Therapeutic nuclear medicine methods encompass external beam therapy, radionuclide delivery, and brachytherapy. Radiation therapy employs X-rays, gamma rays, heavy ions, neutrons, protons, and various other forms of radiation to eliminate cancer cells and remove tumors. Radiation is produced by external beams or radioactive substances positioned within the body adjacent to cancer cells. Systemic radiation therapy employs a radioactive source that circulates through the bloodstream to target specific tissues. Cancer detection and treatment are enhanced through several methods, including imaging and diagnostic nuclear medicine techniques, tailored tumor therapies, external beam radiation therapy, and the administration of radioactive substances to the tumor location. Radiation therapy techniques can be implemented with other treatment modalities to augment the comprehensive treatment program. The primary categories of nuclear medicine encompass general methodologies for imaging the body or selectively administering an absorbed dose to the tumor location. Techniques employed to improve the administration of absorbed dosage to targeted anatomical sites comprise image-guided radiation therapy (IGRT) (Figure 3.6), stereotactic radiosurgery (SRS) (Figure 3.7), stereotactic body radiotherapy (SBRT), intensity-modulated radiation therapy (IMRT), and radioimmunotherapy (RIT) (De Los Santos et al., 2013). Ionizing radiation is helpful for noninvasive imaging techniques that yield information regarding bodily functions and pathological or injured tissues. Low-energy X-rays serve as a fundamental diagnostic tool.

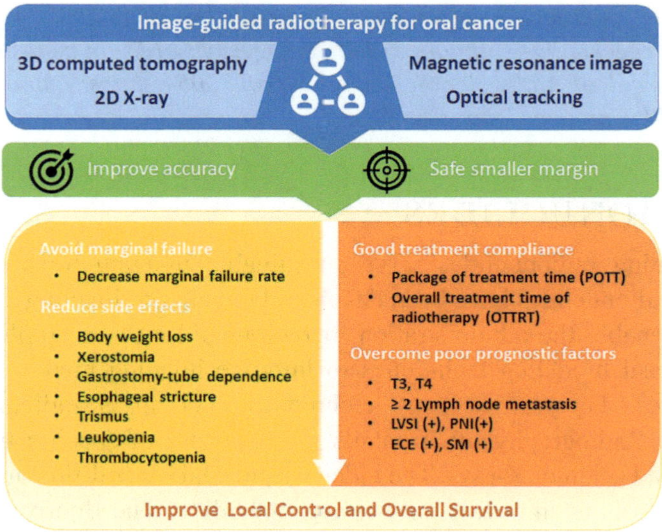

Figure 3.6. *Image-guided radiation therapy (IGRT).*

Source: Nien, H. H., Wang, L. Y., Liao, L. J., Lin, P. Y., Wu, C. Y., Shueng, P.-W., Chung, C. S., Lo, W. C., Lin, S. C., & Hsieh, C. H. (2022). Advances in image-guided radiotherapy in the treatment of oral cavity cancer. Cancers, 14(19), 4630. https://doi.org/10.3390/cancers14194630, Creative Commons License.

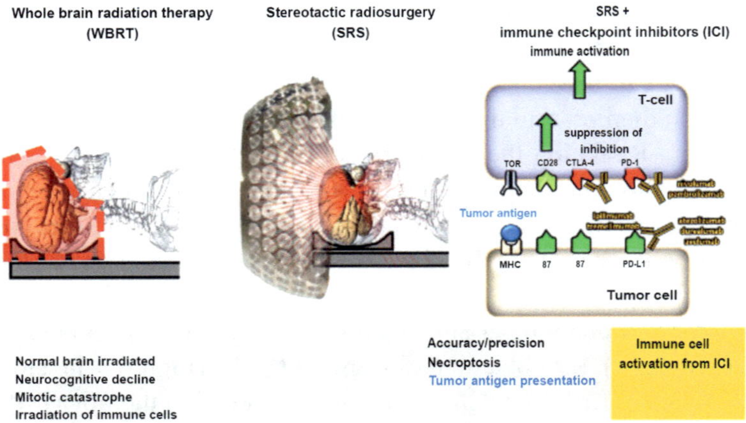

Figure 3.7. *Stereotactic radiosurgery (SRS).*

Source: Lehrer, E. J., McGee, H. M., Peterson, J. L., Vallow, L., Ruiz-Garcia, H., Zaorsky, N. G., Sharma, S., & Trifiletti, D. M. (2018). Stereotactic radiosurgery and immune checkpoint inhibitors in the management of brain metastases. International Journal of Molecular Sciences, 19(10), 3054. https://doi.org/10.3390/ijms19103054, Creative Commons License.

3.4. RADIONUCLIDES

Photon-emitting radionuclides serve as valuable imaging tools, and many diagnostic radionuclides have been developed to enhance imaging capabilities. Computers enable the administration of radionuclides to be employed along with numerical modeling techniques to improve imaging methods and their functionalities. Diagnostic X-ray techniques encompass radiography and fluoroscopy. Radiography often entails the exposure of specific tissues to a wider or broad beam of X-rays. The video output enables real-time imaging that aids in diagnosis or improves an ongoing procedure. The fluoroscopy output may be documented for future review and assessment. Fluoroscopy treatments offer enhanced flexibility and augment the information derived from static X-ray pictures. However, patient doses from fluoroscopy typically surpass those from X-ray imaging (Figure 3.8). The majority of diagnostic radionuclides are photon emitters characterized by shorter half-lives. The objective of nuclear imaging is to get information while administering the least possible absorbed dose to the patient.

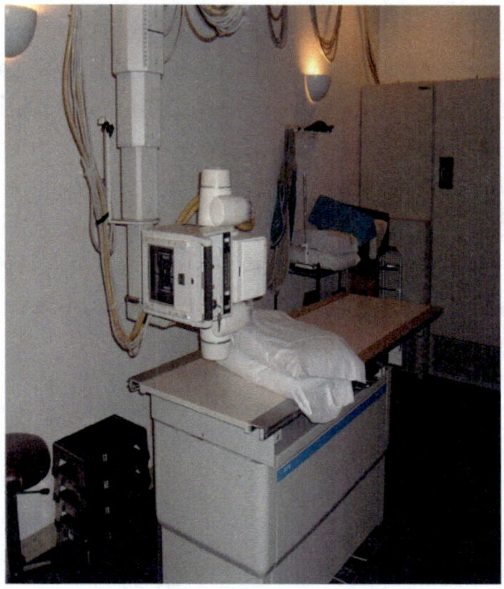

Figure 3.8. *X-ray imaging.*

Source: https://upload.wikimedia.org/wikipedia/commons/7/77/X-ray_table.JPG, Creative Commons License.

Radionuclides are employed to identify the presence and location of infections, pulmonary emboli, myocardial infarctions, and neoplasms. These approaches offer insights for evaluating organ function without disrupting the tissue. This is achieved by adding a photon-emitting radionuclide to a physiologically active chemical to create a radiopharmaceutical.

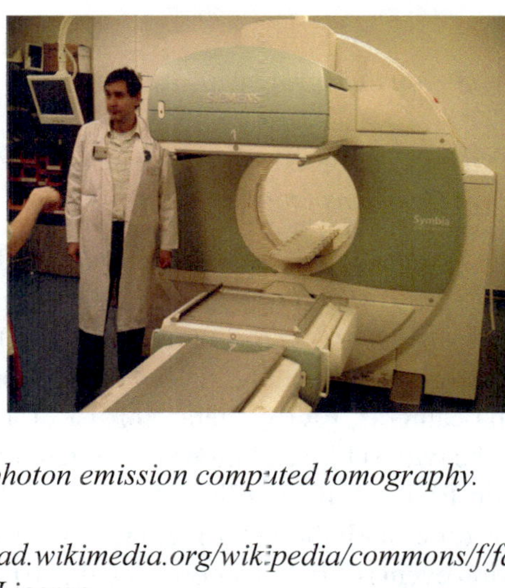

Figure 3.9. *Single-photon emission computed tomography.*

Source: https://upload.wikimedia.org/wikipedia/commons/f/fa/TEMP-Siemens.jpg, Creative Commons License.

The molecule enables the deposition of the radiopharmaceutical at the target site. The radioiodine molecules are selectively absorbed by the thyroid, so enhancing the imaging of that organ. Post-administration, the radiopharmaceutical is assessed utilizing an appropriate measurement instrument like gamma camera to identify photons emanating from the organ. Diagnostic procedures involve imaging the cross-section (body) by tomography. Tomography exceeds planar imaging due to its capability to visualize underlying organs through computational methods. The approach acquires several axial slices of an area.

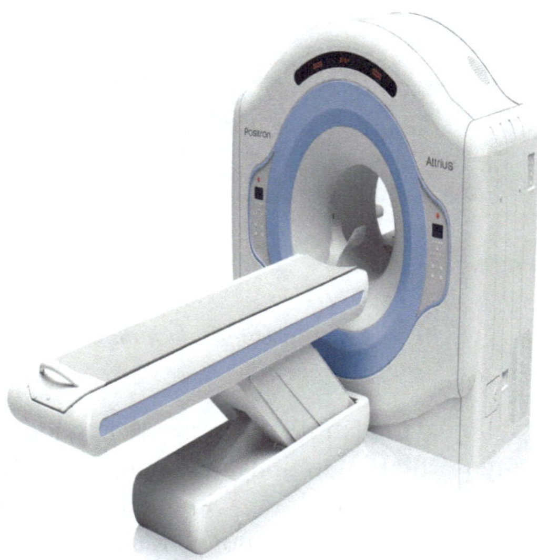

Figure 3.10. *Positron emission tomography.*

Source: https://upload.wikimedia.org/wikipedia/commons/e/e6/Positron_emission_tomography.png, Creative Commons License.

Computed tomography primarily exists in two forms: single-photon emission computed tomography (SPECT) (Figure 3.9) and positron emission tomography (PET) (Figure 3.10) (O'Connor and Kemp, 2006). The SPECT approach employs a revolving gamma camera and radionuclides used in traditional planar imaging. Unlike rotating gamma cameras, PET cameras consist of a stationary array of equipment that detect the two 511 keV photons generated by the annihilation of a positron-electron pair. PET scanners utilize coincidence detectors positioned 180 degrees apart or time-of-flight (TOF) circuits to ascertain the spatial coordinates (annihilation event) (Krishnamoorthy et al., 2017). These events are compiled to generate an image of the scanned organs. PET procedures employ positron-emitting radionuclides such as ^{11}C, ^{13}N, ^{15}O, and ^{18}F. The released positrons engage with electrons in the target tissue to generate annihilation photons. Short-lived positron-emitting radionuclides must be generated within the medical facility likely via a cyclotron, whereas ^{18}F is generally transported to the institution from a supplier. Radiation therapy employs radioactive substances or diverse radiation beams to deliver energy to the body, aiming to eliminate damaged tissue. These therapeutic modalities employ beta- and alpha-emitting radionuclides, proton, electron, and heavy ion beams, brachytherapy, and radioimmunotherapy (RIT) (Figure 3.11) to administer absorbed doses to the tumor or affected tissue site. Therapeutic radionuclides are engineered to provide substantial energy to a specific tissue. They integrate beta- and alpha-emitting radionuclides into the radiopharmaceutical.

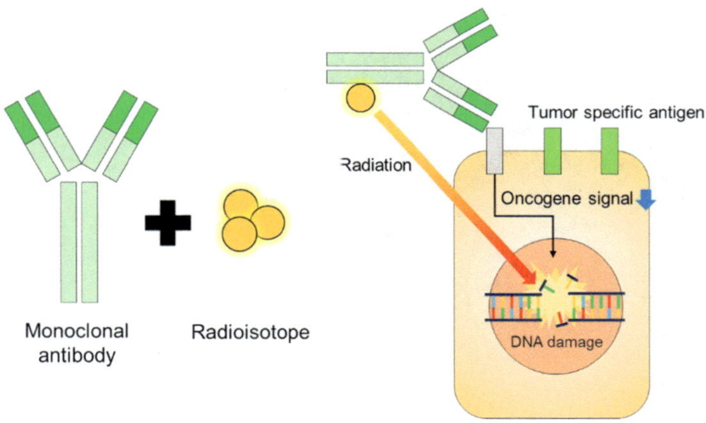

Figure 3.11. *Radioimmunotherapy.*

Source: Zaheer, J., Kim, H., Lee, Y. J., Kim, J. S., & Lim, S. M. (2019). Combination radioimmunotherapy strategies for solid tumors. International Journal of Molecular Sciences, 20(22), 5579. https://doi.org/10.3390/ijms20225579, Creative Commons License.

3.5. THERAPEUTIC RADIONUCLIDES

Therapeutic radionuclides serve multiple purposes, including the management of hyperthyroidism, (Figure 3.12) alleviation of pain related to metastatic cancer, and therapy of various malignancies and tumors (Elgazzar and Sarikaya, 2022). The beta-emitting radiopharmaceutical destroys a tumor upon local deposition; however, the radiation may also impact healthy tissue if its range exceeds the tumor margin. High-energy beta radiation and bremsstrahlung from nuclides such as ^{32}P and ^{90}Y frequently have a range that exceeds the target tissue. The irradiation of healthy tissue adversely affects patient recovery and may influence their quality of life. The use of alpha-emitting radionuclides in radiopharmaceuticals minimizes the irradiation of healthy tissue. Alpha particles have a shorter range in tissue compared to high-energy beta particles, and their application offers distinct advantages for the preferred delivery of absorbed doses to cancer cells. The utilization of alpha-emitting radionuclides in therapy is presently constrained by their availability and the economic feasibility of their production.

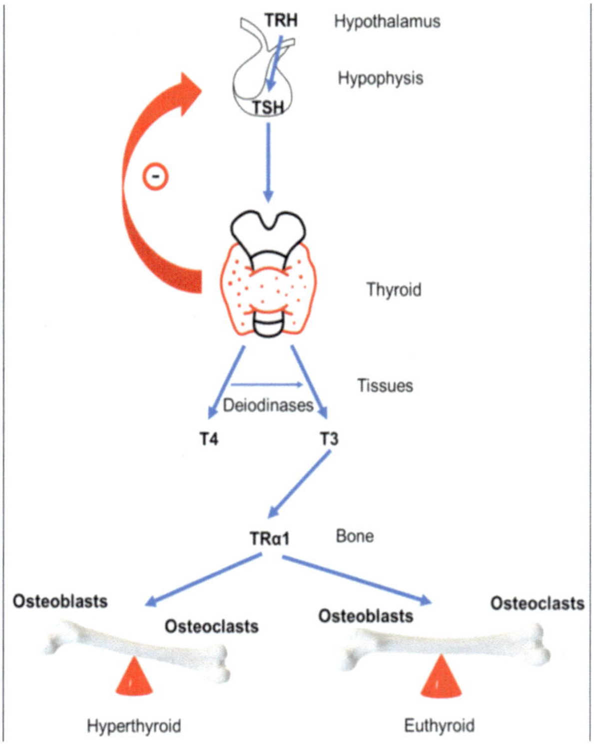

Figure 3.12. *Hyperthyroidism.*

Source: Mudri, D., Bilić Ćurčić, I., Meštrović, L., Mihaljević, I., & Kizivat, T. (2023). Hyperthyroidism and Wnt signaling pathway: influence on bone remodeling. Metabolites, 13(2), 241. https://doi.org/10.3390/metabo13020241, Creative Commons License.

High therapeutic dosages are administered via accelerators and teletherapy equipment, including ^{60}Co systems. External beams emit diverse radiation modalities that either directly target the tumor with neutrons, electrons, photons, protons, and heavy ions or engage with a substance integrated within the tumor volume. Beam therapy must also reduce the absorbed dose in the tissue between the skin and the tumor mass. There is considerable variability in penetration depths and energy deposition attributes according to the kind of radiation and its energy level. Divergence and attenuation of photon beams diminish photon fluence as a function of tissue depth. The electron density reaches an equilibrium value as it passes through tissue. The interplay of these factors generates an absorbed dose curve that ascends to a peak and thereafter declines with greater tissue depth. Electron backscatter elevates the surface absorbed dose to a range of 15% to 100% of the maximal dosage. The depth of the maximum absorbed dose escalates with elevated beam energy. The absorbed dose curve for 15 MeV photons reaches its peak at around 2.7 cm depth, with therapeutically relevant radiation extending to roughly 8 cm tissue depth. Electron beams cause primary electrons to dissipate energy in tissue, generating elevated ionization densities per unit length as they attain their maximum range. For tissue depths exceeding the maximum range, the absorbed dose of electrons diminishes swiftly to a minimal fraction of the maximum absorbed dose.

For electron energies under 1 MeV, the peak absorbed dosage is located near the skin surface. Given that the majority of lesions reside beneath the skin surface, employing higher-energy electron beams is beneficial for achieving the necessary tissue depth. By appropriately selecting the beam energy, the tumor is targeted while preserving the surrounding healthy tissue. A chest wall tumor must be treated without compromising the underlying lung tissue. As the energy of the electron beam escalates from 4 to 20 MeV, the configuration of the absorbed dose curve transitions from a surface peak to a more extensive plateau penetrating deeper into the tissue. Beyond 20 MeV, the plateau extends, resulting in the loss of the benefit of preserving healthy tissue at depth. The effective electron energy range is between 4 and 20 MeV (Schardt et al., 2010). Proton beams provide a consistently low absorbed dosage that ends in a sharp peak at the end of the absorbed dose curve. The absorbed dose is highly localized, resulting in significant energy deposition in tumors while minimizing doses to healthy tissue. The capacity to monitor a typical beam is constrained. Monitoring is essential to guarantee the complete tumor volume is subjected to irradiation. Brachytherapy entails the insertion of radioactive substances in immediate proximity to the tumor mass. It is utilized for malignancies that are reachable via natural body cavities or in proximity to body surfaces. An optimal brachytherapy radionuclide possesses a range that restricts its energy deposition to the tumor volume.

3.6. BRACHYTHERAPY

Conventional brachytherapy sources utilize radionuclides that emit beta particles, photons, and conversion electrons. Brachytherapy may be utilized independently or in conjunction with other treatments, including as external beam radiotherapy, and chemotherapy. The radionuclides commonly utilized in brachytherapy, comprise of ^{125}I, ^{137}Cs, ^{192}Ir, and ^{198}Au (Coursey and Nath, 2000). The brachytherapy source material is contained within a protective capsule, permitting the escape of ionizing radiation to irradiate adjacent tissue but preventing the dissolution of the radioisotope in bodily fluids and its translocation to other tissues. Brachytherapy sources are either extracted post-treatment or, in the case of certain short-lived radionuclides, let to remain in situ. The brachytherapy source exclusively irradiates a confined tissue volume, hence minimizing the absorbed dose to healthy tissues.

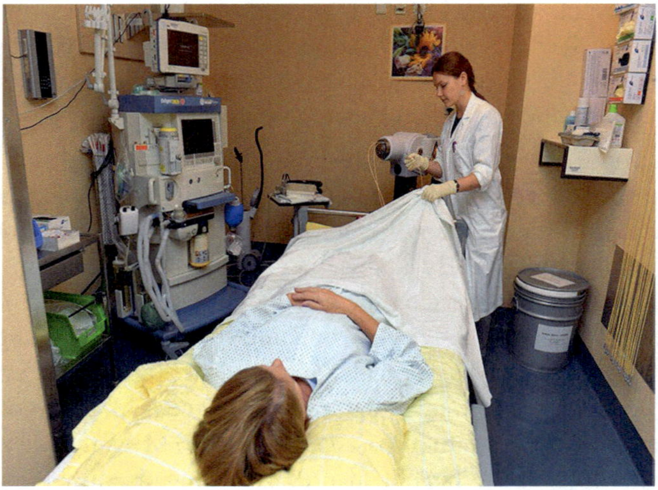

Figure 3.13. *Brachytherapy.*

Source: https://www.iaea.org/sites/default/files/styles/original_image_size/public/brachytherapy_1.jpg?itok=sSVG80EZ, Creative Commons License.

Brachytherapy (Figure 3.13) has the capability for more localized dosage administration. A brachytherapy treatment procedure typically requires less time than other radiotherapy methods and can often be conducted on an outpatient basis. Treatment outcomes indicate that the success rates of brachytherapy are analogous to those of surgery and external beam radiation therapy. The outcomes tend to enhance when multimodal techniques are employed. Brachytherapy exhibits a comparatively low occurrence of unfavorable side effects. These adverse consequences impact quality of life. Brachytherapy is frequently employed to address malignancies of the breast, cervix, prostate, and skin. It is also employed to address malignancies in various body tissues, including the female reproductive system, gall bladder, anus, brain, digestive tract, esophagus, eye, head, neck, lip, male reproductive system, mouth, nasopharynx, oropharynx, rectum, respiratory system, and urinary system. Brachytherapy sources are often characterized by their administered absorbed dose rates. Low-dose rate (LDR) brachytherapy sources generate an absorbed dose rate of up to 2 Gy/h (Nag et al., 2002). LDR brachytherapy is frequently employed for malignancies of the oral cavity, oropharynx, prostate, and other sarcomas. Medium-dose rate (MDR) brachytherapy is defined by absorbed dose rates ranging from 2 to 12 Gy/h (F. D. Patel et al., 1998). High-dose rate (HDR) brachytherapy administers an absorbed radiation rate above 12 Gy/h (Nag et al., 2000). The prevalent applications of HDR brachytherapy are in malignancies of the cervix, esophagus, breast, lungs, and prostate. Brachytherapy may also utilize pulsed sources. Pulsed dose rate (PDR) brachytherapy employs short radiation pulses to replicate the cumulative rate and efficacy of low dose rate (LDR) treatment. The PDR source is generally a ^{192}Ir source, approximately 4×10^4 MBq, which is operated similarly to a HDR system. PDR brachytherapy is utilized for gynecological, head and neck malignancies. The standard PDR methodology exceeds a prolonged LDR interstitial therapy spanning several days, incorporating a sequence of short HDR irradiations at a specified interval. PDR is intended to utilize computer-controlled remote technology in combination with brachytherapy implant catheters. The residence periods in each position are modified to achieve the necessary absorbed dose. The PDR method offers several benefits. The patient is exposed to radiation for only a fraction of the time. The absorbed dosage administered to the tumor is maximized through computer control. NCRP 155 establishes a distinction between sealed and unsealed radioactive sources. All unsealed sources are classified as radiopharmaceutical therapy sources. Many sealed sources are classified as brachytherapy sources. Most of these brachytherapy sources employ photon, beta, and conversion electron sources. The application of low-energy photons, low-energy beta particles, and low-energy conversion electrons increases energy deposition inside the tumor volume while reducing the dose to healthy tissue. The tumor dose is optimized through the utilization of alpha-emitting radioactive substances. The limited range of alpha particles optimizes the absorbed dose administered to the tumor. Along with alpha particles, the application of neutron radiation in brachytherapy is now under development. Californium-252 has a half-life of 2.646 years, decaying predominantly by alpha emission, although spontaneous fission occurs with an approximate yield of 3.1%. It is the sole neutron-emitting radioisotope employed in radiotherapy applications. Along with spontaneous fission neutrons and alpha particles, the decay products of ^{252}Cf produce photons and beta particles. The application of a ^{252}Cf brachytherapy source for tumor treatment predominantly involves neutrons and photons. The alpha and beta particles emitted by ^{252}Cf have exceedingly limited ranges in biological tissue. Approximately one-third of the radiation dose absorbed by tissue surrounding a ^{252}Cf brachytherapy source is attributable to gamma rays, whereas two-thirds is due to neutrons. ^{252}Cf brachytherapy has been utilized in clinical studies for several cancer types, including cervical, esophagus, lip, oral cavity, rectal, skin, tongue, and soft tissue cancers. It has additionally been utilized in combination with BNCT.

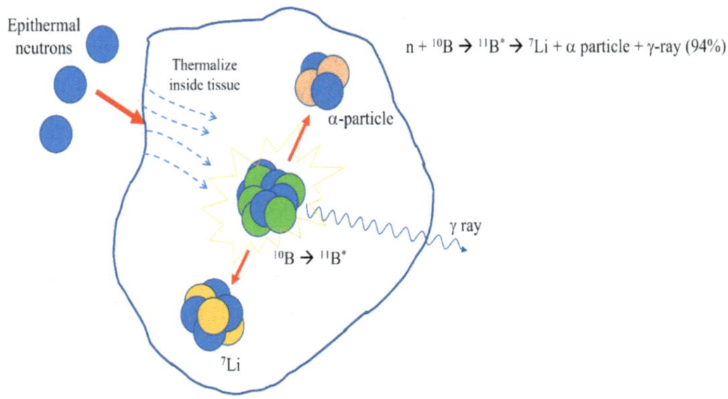

Figure 3.14. *Boron-enhanced neutron brachytherapy.*

Source: Monti Hughes, A., & Hu, N. (2023). Optimizing boron neutron capture therapy (BNCT) to treat cancer: An updated review on the latest developments on boron compounds and strategies. Cancers, 15(16), 4091. https://doi.org/10.3390/cancers15164091, Creative Commons License.

Boron-enhanced neutron brachytherapy (BENBT) (Figure 3.14) integrates neutron brachytherapy with boron neutron capture therapy (BNCT). In BNCT, a boron-containing molecule is delivered and localized to the tumor location. The tumor is subsequently bombarded with thermal neutrons, resulting in the production of heavy ions by the ^{10}B(n, □)^{7}Li process (Zaidi et al., 2018). The heavy ions produced by the reaction have a limited range and are efficient in delivering absorbed dose within the tumor volume. A fraction of the fast neutrons generated by the spontaneous fission of ^{252}Cf are attenuated to thermal energy when they cross the tumor. BENBT or Boron-enhanced neutron brachytherapy administers absorbed dosage to the tumor via direct 252Cf neutron radiation, supplemented by extra heavy ion absorbed dose from the ^{10}B(n, □)^{7}Li reaction. Alternative brachytherapy methods include the permanent insertion of radioactive substances embedded in glass and resin microspheres, as well as low-energy applications. RIT or Radioimmunotherapy constitutes a fusion of radiation and immunotherapy. Immunotherapy employs a designed chemical, specifically a monoclonal antibody, to attach to cancer cells. Monoclonal antibodies operate similarly to natural antibodies that target viruses. At RIT, a monoclonal antibody is conjugated with radioactive material. Upon injection, the monoclonal antibody attaches to the cancer cell and administers substantial absorbed dosages directly to the tumor

3.7. RADIOIMMUNOTHERAPY (RIT)

Radionuclides such as ^{90}Y and ^{131}I are utilized in RIT. RIT is presently employed in the treatment of multiple types of lymphoma. This method has the capacity to address various disease types, including brain tumors, ovarian cancer, colon cancer, leukemia, melanoma, and prostate cancer. Imaging supporting RIT is frequently conducted with a gamma camera. Low-energy photon sources provide the capability to treat malignancies via catheters. A low-energy photon emitter possesses a limited range compared to a higher-energy photon and enables extended implantation durations. ^{169}Yb, possessing a half-life of 32.03 days, emits photons, with peak yields occurring beneath 200 keV (Patel et al., 2001). It also emits electrons, with the predominant yields exhibiting energies below 6 keV. Due to these features, nuclides like ^{169}Yb provide alternative cancer treatment alternatives with prolonged treatment durations without requiring highly shielded environments. Low-energy brachytherapy provides the capability for targeted energy delivery for specialized therapeutic uses. These applications involve the disruption of a tumor blood supply through the destruction of its vascular architecture. Tumor vascular disruption has been achieved using chemotherapy and radiotherapy.

3.8. THERAPEUTIC APPLICATIONS

Miniature X-ray sources also offer low-energy photons for therapeutic uses. Therapeutic uses encompass interstitial radiosurgery and intravascular irradiation to avert restenosis. Standard voltages typically vary from 20 to 40 kV. This approach has supplementary therapeutic implications due to the significant biological efficacy of low-energy photons relative to higher-energy gamma rays. This emerges due to the photoelectric absorption at low energy. IGRT employs imaging tools during radiation therapy to accurately target the delivery of absorbed dosage to the tumor location. In IGRT, the apparatus administering the absorbed dose incorporates imaging technologies to visualize the tumor during the treatment process. The imaging apparatus enables essential modifications to accurately focus on the tumor while sparing surrounding healthy tissue from radiation exposure. Contemporary IGRT techniques utilize light-emitting diode (LED) emissions aimed at the body. IGRT is frequently employed to address malignancies in mobile tissues, such as the lungs, liver, and prostate, during nuclear medicine procedures. The IGRT methodology is employed together with IMRT (Image-modulated Radiation Therapy), external beam treatment, SRS, and SBRT. IMRT is an external beam therapy technique that uses a linear accelerator to produce photons with energy approximately 10 MeV (Greene and Williams, 2017). This approach accurately administers absorbed dose to a tumor or designated regions within the tumor. Computed tomography or MRI is frequently employed to ascertain the absorbed dose profile that aligns with the tumor morphology. Intravenous contrast chemicals may be administered during imaging to enhance the delineation of tumor morphology. The combinations of various intensity-modulated fields from varied beam orientations produce an optimal absorbed dose profile that maximizes tumor dosage. The IMRT minimizes the absorbed radiation to healthy tissue, allowing for larger doses to be administered to malignancies with reduced side effects compared to conventional radiotherapy treatments. Due to its complexity, IMRT necessitates marginally extended durations and planning compared to other radiations. IMRT is predominantly employed in the treatment of central nervous system disorders, head and neck neoplasms, and prostate carcinoma. IMRT has been utilized to treat cancers of the lung, gastrointestinal tract, breast, and thyroid. IMRT frequently necessitates several treatment sessions. The total number of IMRT sessions and the corresponding absorbed dose are contingent upon the nature, location, and size; the doses to surrounding tissues; and the overall health status. Patients are often scheduled for IMRT sessions lasting 10 and 30 minutes, five days a week for 5 to 8 weeks. Stereotactic radiosurgery (SRS) is a precise mode of radiotherapy originally employed to address tumors within the brain. The methodologies established with SRS are also being utilized to address other malignancies using a process known as stereotactic body radiotherapy (SBRT). SRS and SBRT provide a non-invasive alternative to surgical intervention. SRS is a non-invasive method that accurately administers absorbed doses at significantly elevated levels compared to conventional radiation therapy. This accuracy reduces the exposure to surrounding healthy tissue. SRS and SBRT utilize three-dimensional imaging and localization methods to provide precisely targeted gamma-ray or X-ray beams that converge on a tumor. This method can also be utilized along with IGRT to enhance treatment accuracy.

3.9. IMAGING TECHNIQUES

Three-dimensional imaging techniques, such as CT, MRI, and PET/CT, are employed to identify the tumor within the body and delineate its precise geometry

(Schmidt et al., 2006). The imaging data aid in delineating the orientation of the photon beams that converge on the tumor. SRS is usually performed in a single day; however, for tumors exceeding 1 inch in diameter, two to five sessions are sometimes employed. SBRT addresses many malignancies located in the lung, neck, head, liver, prostate, and spine. Three types of machines are employed to conduct SBRT. A gamma knife utilizes concentrated beams of gamma rays directed at the target area and is commonly employed to treat brain malignancies. Linear accelerators provide high-energy photons for the treatment of large malignancies. Security procedures are necessary to guarantee that therapeutic sources are employed for their designated purpose. 10CFR37 was instituted to supersede the directives and to implement restrictions concerning physical protection measures, fingerprinting, and background checks for any licensee possessing a Category 1 or Category 2 quantity of radioactive material (Saha, 2018). The stipulations for accessing radioactive material, as well as its utilization, transfer, and transportation, are delineated in 10CFR37. This regulation impacts medical establishments that employ radioactive materials.

3.10. RADIATION EXPOSURE

Acute adverse effects manifest during or shortly following radiation exposure and generally persist for many weeks. The adverse effects include fatigue and dermatological disorders characterized by dryness, irritation, erythema, excess sensitivity, and edema. Additional effects include diarrhea, dysphagia, digestive complications, nausea and emesis, and edema in the treatment region, as well as urinary and bladder disorders. Late side effects, manifesting months or years post-treatment, occur less frequently than early side effects and are often permanent conditions. Alterations in tissue within the brain, colon, lung, kidney, joints, mouth, rectum, and spinal cord are prevalent late consequences. Other late effects include cardiovascular disease (CVD), infertility, and secondary malignancies. Radiation constitutes a crucial element in effective cancer treatment, with an estimated 50% of all patients having undergone radiation therapy. Many patients experience late consequences generated by radiation subsequent to cancer survival. Secondary primary cancer (SPC) and CVD are two of the prevalent and significant adverse effects linked to radiation therapy. The risk of SPC subsequent to radiation therapy is substantiated by evidence. According to NCRP 170, quantitative risk estimates for cardiovascular disease are not yet definitively established (Calabrese, 2022). Radiobiology establishes a basis for comprehending the essential cellular damage pathways that result in the occurrence of SPC and CVD. Genetic factors are recognized to influence individuals to the development of several primary malignancies. These genetic conditions elucidate the interplay between high-dose radiation and genetic tendencies in the etiology of SPCs. Conventional radiation therapy delivery methods include beam therapy utilizing various radiation types, ^{60}Co teletherapy equipment, and brachytherapy. NCRP 160 indicates that contemporary radiation therapy methods have evolved in recent decades, resulting in increasing patient doses. The elevated doses and SPC and CVD effects indicate that the advantages of radiation therapy must be evaluated against the hazards. The information in NCRP 160 and 170 should aid clinicians in identifying the optimal treatment strategy. Heavy ions like ^{12}C demonstrate a low relative biological effectiveness (RBE) at the body surface, facilitating a favorable probability for healthy tissue recovery (Karger et al., 2021). A higher RBE is observed where the damaged DNA in this area is more challenging to repair. As the mass of heavy ions exceeds that of carbon, elevated RBE leads to a greater absorbed dose in healthy tissue near the entrance region or area. The entrance RBE escalates with the atomic number of the heavy ion beam. The RBE influences the efficacy of DNA damage repair, making the repair capacity of irradiated tissue significant. The repair of healthy tissue is a crucial factor for slowly proliferating cancers that possess considerable repairing ability and are typically resistant to radiation. Multimodal therapy integrates two or more diverse modalities to address a disease. These therapeutic modalities typically include surgical intervention, chemotherapy, and radiotherapy. Similar to radiotherapy, chemotherapy impacts healthy tissue, potentially resulting in adverse effects such as hair loss and nausea. The choice of a certain combination of strategies is based on the patient health, the physician expertise and success with a particular protocol, and the characteristics of the disease being addressed. The adverse effects of prostate cancer treatment include erectile dysfunction, incontinence, and extended healing period. Tracking methodologies are based on the dosage administration technique. In the context of a radiopharmaceutical, tracking becomes a secondary concern when the molecular species is predominantly localized within the tumor limits. Localizing the dose within the tumor boundary requires the use of a short-range radiation type. These radiation modalities enable the targeted deposition of absorbed dosage within the tumor volume.

3.11. BEAM THERAPY

Identifying the location of absorbed dose is critical when employing external beam therapy. The beam can be monitored under specific conditions. A heavy ion loses neutrons when it crosses tissue. Removing neutrons from the nucleus elevates the proton-to-neutron ratio, so distancing the nucleus from the line of stability. The presence of surplus protons in the nucleus promotes positron emission and electron capture processes. Positron emission generates two annihilation photons, which are detected with PET (Abuelhia et al., 2007).

Photons diminish in intensity exponentially as they traverse tissue, whereas charged particles primarily deposit their energy at the Bragg peak towards the end of their range. The Bragg peak has a spike for protons and heavy ions, while it is broader for charged pions. Antiprotons interact with protons in tissue, resulting in the antiproton energy deposition curve reflecting a synthesis of the Bragg peaks for these radiation types and their annihilation products. A heavy ion is typically characterized as a nucleus with a mass greater than that of a proton. Conventional beam radiotherapy utilizing photons and electrons is constrained due to the irradiation of healthy tissue during treatment. To overcome the physical and biological constraints of conventional radiation, researchers proposed the utilization of heavy ions. Heavy ions enable the delivery of a larger absorbed dose to a deep-seated tumor compared to proton treatment. This occurs because heavy ions have an inverse absorbed dose profile, with the dosage increasing as penetration depth increases. The maximum absorbed dose is attained during the Bragg peak. The energy deposition from a photon, electron, muon, pion, proton, or heavy ion at a tumor site is often 100 MeV or even less (Nunes, 2014). While this energy deposition is beneficial in therapeutic applications, greater energy depositions can be achieved by annihilation events involving hadrons. An implementation of this principle is antiproton treatment, which involves the annihilation of a proton and an antiproton. The mass conversion of a proton-antiproton pair during annihilation generates a high-energy deposition source that may have potential implications in cancer therapy. Antiprotons, akin to protons and heavy ions, primarily deposit their kinetic energy at the end point of their range, known as the Bragg peak. The antiproton annihilates upon colliding with a proton, releasing energy. Upon annihilation, additional particles and photons are generated, depositing a fraction of their energy at the tumor location. A significant portion of the accessible annihilation energy of 1.88 GeV is dissipated at the tumor site as charged pions, neutrons, and photons exit the target volume. Preliminary measurements at CERN estimate that the dose deposition from antiprotons is similar to that of protons; however, Bragg antiprotons deposit an additional 30 MeV within a few millimeters of the annihilation event. The local energy deposited effectively doubles the absorbed dosage per particle in the Bragg peak in comparison to protons. The Relative Biological Effectiveness (RBE) of the particles contributing to the dosage exceeds that of protons. A segment of the residual annihilation energy departing from the tumor site and entering the body may be utilized for real-time imaging of the dose distribution. The resultant antiproton depth dose curve can be compared with that of protons assessed under analogous conditions. Proton-antiproton annihilation events yield an average of 4 to 5 charged and neutral pions with a mean energy of around 400 MeV, in addition to 3 high-energy gamma rays (Hori and Walz, 2013). The photons present the capability to monitor the beam and enhance energy deposition through computed tomography methodologies. Beta-emitting radionuclides are frequently utilized in medical therapeutic applications. Beta radiation eliminates malignancies; nevertheless, its range frequently exceeds the tumor location. An alternative to beta emitters is the utilization of alpha-emitting radionuclides.

3.12. ALPHA THERAPY

Targeted alpha treatment provides substantial advantages for both patients and hospital personnel. The application of targeted alpha therapy is constrained by the accessibility of suitable radionuclides and the capacity to manufacture them economically.

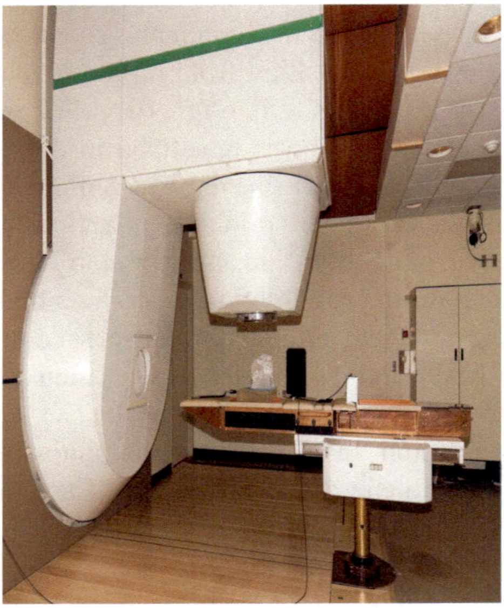

Figure 3.15. *Neutron therapy.*

Source: https://upload.wikimedia.org/wikipedia/commons/5/5e/UW_Therapy_room.jpg, Creative Commons License.

The utilization of alpha-emitting radionuclides presents the possibility of internal exposure to these substances. Suitable contamination management measures are necessary to reduce surface contamination and prevent internal exposures. The fabrication of these materials by accelerator presents the possibility of external exposures from accelerator beams or their resultant reaction products. These nuclides are constituents of a decay chain and can be obtained from a source of the parent nuclide. ^{213}Bi and ^{225}Ac are from ^{233}U and have been utilized in the treatment of leukemia and solid tumors, respectively (Morgenstern et al., 2020). Radium has therapeutic potential for alleviating bone cancer pain. This radionuclide is generated in an accelerator and can be isolated from the target substance by chemical separation methods. Neutrons have been utilized in therapeutic applications due to their higher relative biological effectiveness (RBE) compared to protons, pions, muons, photons, and electrons. While neutron therapy results in substantial dose deposition, this advantageous aspect must be weighed against the adverse effects stemming from inadequate dose localization.

3.13. NEUTRON THERAPY

Neutron therapy (Figure 3.15) is conducted using various energy levels, including thermal, epithermal, and other regimes. Neutron capture therapy utilizes the thermal and epithermal energy domains, while fast neutron therapy (FNT) employs the highest energy spectrum. In neutron capture therapy, an isotope with a significant absorption cross-section for thermal or epithermal neutrons is administered into the body, typically via intravenous injection. The isotope is included into a molecule engineered to concentrate in certain tissue. Neutron capture therapy is employed to activate ^{10}B-labeled compounds that accumulate in glioblastoma tumors. BNCT specifically targets and irradiates tumor cells. Targeted administration of ^{10}B to the tumor location, followed by irradiation with thermal/epithermal neutrons, facilitates dosage localization. The limited range of the $^{10}B(n, \alpha)^{7}Li$ reaction products confines or restricts the majority of the treatment to the boron-loaded tumor cells. FNT employs neutron beams with energy of 15 MeV. The therapeutic effect is attained through the recoil of protons and heavier ions produced by rapid neutron interactions. Fast neutrons are utilized in the treatment of various malignancies, including those located in the head and salivary glands, neck, and soft tissue. Neutrons for BNCT or FNT can be produced via several processes, including $^{2}H(d, n)^{3}He$ and $^{3}H(d, n)^{4}He$ (Verma, 2007). Neutron therapy is very effective for treating inoperable, radioresistant cancers located throughout the body. Fast neutrons possess the ability to manage substantial tumors, as fast neutron therapy is independent of oxygen availability for the eradication of cancer cells. The FNT methodology is unaffected by the stage of cancer cells. Due to their significant relative biological effectiveness (RBE), the neutron absorbed dosage necessary to eliminate cancer cells is approximately one-third of the dose needed for photons, electrons, or protons. A complete neutron therapy is administered in about 10–12 sessions, in contrast to the 30–40 sessions required for low linear energy transfer (LET) radiation. The efficacy of neutron beam therapy varies according on the patient, cancer stage, and histology; however, some malignancies have been addressed via this method. These applications include tumors located in the abdomen and pelvis, chest, bone, cartilage, and soft tissue, head and neck, skin etc. A constraint in standardizing neutron therapy methods is the significant variability in beam parameters, energy spectra, and collimation procedures among neutron facilities. This therapeutic method is restricted to facilities equipped with a neutron radiation source. The ability to generate neutrons is often confined to establishments equipped with a research reactor or accelerator. Potential applications necessitate more investigation and involve inoperable pancreatic and rectal tumors. Conventional radiation frequently entails the placement of the radionuclide within a tumor mass. The radiation emitted from the disintegration of the radionuclide is utilized to deliver absorbed dose to the tumor mass, hence diminishing its size. A substantial absorbed dose can eliminate the tumor. A radionuclide is employed to obliterate the tumor vascular supply, so depriving it of essential nutrition instead of directly targeting the tumor mass. Vascular disruption drugs have been integrated into chemotherapy and radiation protocols. These methodologies are also referred to as antiangiogenic or radioembolization therapy. Radiotherapy methods targeting vascular disruption have been widely employed in liver tumors using ^{90}Y microspheres (Salem et al., 2002).

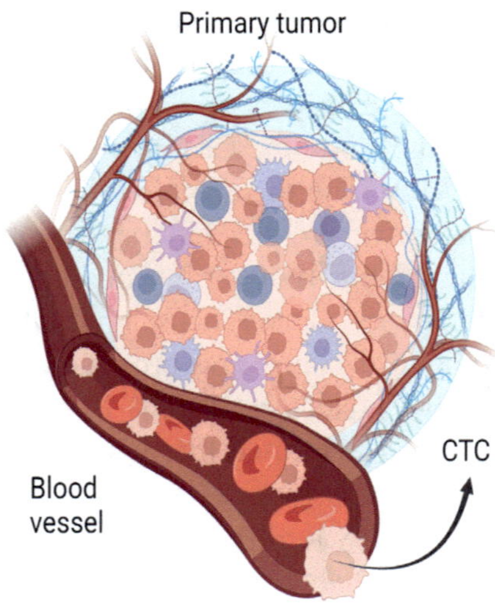

Figure 3.16. *Tumor vascular system.*

Source: https://upload.wikimedia.org/wikipedia/commons/0/06/Circulating_tumor_cells_-_en.png, Creative Commons License.

3.14. TUMOR

In normal tissues, the vascular architecture is organized to ensure optimal nutrition conditions. Expanding tumors possess an irregular vasculature that is neither fully established nor sufficient to optimally sustain the tumor cells. The tumor vascular (Figure 3.16) can be compromised with a suitable drug. Typical anomalies in a tumor vascular architecture involve dilated vessels with elongated shapes and sudden variations in sizes. Blood circulation in these vessels is slow and irregular. This flow pattern provides inadequate nourishment compared to that supplied to normal cells, leading to hypoxic regions typical of solid tumors. Hypoxic conditions diminish the efficacy of both chemotherapy and radiation. The deficiency of oxygen confers a certain level of radioresistance to tumor cells in comparison to oxygenated cells. The growth of a tumor relies on adequate nourishment; hence its survival is compromised by the interruption of blood flow. Obstructing vascular supply provides an alternative method. Cancer is a disease characterized by the unregulated replication of cells. Cancerous cells generate a tumor, although in the initial stages, they do not possess their own blood vessels. At this first phase, cancer cells depend on adjacent tissue for essential nutrients, such as oxygen and glucose. Tumor cells within the interior secrete proteins to indicate nutrient shortage, which then spread until they encounter the blood vessels. This process promotes the formation of new blood vessels that provide the tumor with nutrients necessary for its accelerated growth. Accelerated growth (tumor) generates blood vessels that are irregular and have gaps than those of healthy vessels. The gaps fluctuate and generally range from several 100 nanometers to a few micrometers, exceeding the 2–6 nm pore size found in standard blood arteries. Nanoparticles measuring 10–300 nm can cross the tumor blood vessels but do not infiltrate healthy tissue (Jiang et al., 2015). Loading nanoparticles with radioactive substances or chemotherapeutic agents enables the targeted damage of malignant tissue while preserving healthy tissue.

3.15. NANOPARTICLES

Nanoparticles aggregate within the tumor mass; yet, competing biological processes may restrict their assimilation into the tumor. The microsphere dimensions range from 20 to 60 μm, whereas the nanoparticles range from 10 to 100 nm. The primary factor restricting nanoparticle integration into a tumor mass is the human immune system, which responds to nanoparticles and viruses in a similar way. Phagocytes target both nanoparticles and viruses in a manner similar to the body's immune response to bacteria. The phagocytes engage with the nanoparticles and are then transferred to the liver and spleen. This action inhibits their effectiveness in targeting the tumor and administers the absorbed dose to these structures. A secondary impact is the elevated fluid pressure that arises within tumor cores. In healthy tissue, fluid erupts from blood vessels into adjacent tissue, where it is collected by the lymphatic system and subsequently returned to the bloodstream. To realize the full potential of nanotechnology, it is essential to overcome the previously identified consequences. There are methods to overcome these obstacles. Antiangiogenic drugs reduce pressure within the tumor core, hence enhancing nanoparticle infiltration. Multistage nanoparticles are currently being studied. These particles integrate the larger particle capacity to aggregate within the tumor mass with the smaller particle capability to infiltrate malignant tissue. Engineering nanoparticles to leak from the tumor vasculature into the tumor mass constitutes merely a preliminary phase in tumor elimination (Wang et al., 2023). The efficacy of nanoparticle localization within a tumor mass is influenced by the physical characteristics of the particles, their surrounding environment, and the interaction effects between the particles and their environment. These impacts include fluid dynamics within tumor blood vessels, stresses on blood vessel walls, nanoparticle velocity, and local tissue permeability. Many clinical trials are investigating the application of nanoparticles in cancer treatment. Many applications utilize nanoparticles for the delivery of established chemotherapeutic drugs. Others employ novel chemical agents and multimodal therapy incorporating radioactive materials. Significant research is still required to refine and enhance nanoparticles as a validated cancer treatment method. Nanoparticles are engineered with a particular composition and structure. These particles offer adequate surface area and structure to enable the adhesion of radioactive or chemical agents to a tumor location. The transportation and administration of therapeutic radionuclides are enhanced by coating polymers with reactive functional groups of the radioactive substance. The polymer coatings are engineered to diminish the absorption of radioactive substances in healthy tissue and to prolong their circulation period in the bloodstream. Surface shape, particle size, and surface charge are critical attributes that influence the biodistribution of nanoparticles. These attributes are modified to enhance the transport of the nanoparticles to the designated location. Active and passive targeting methodologies are employed to improve the delivery of nanoparticles to tumors. Through passive targeting, the nanoparticles arrive at the tumor site via their permeable vascular architecture and persist due to restricted or constricted lymphatic drainage channels. In an active mode, nanoparticles are engineered to selectively target certain tumor masses. The disparities in the properties of tumor-specific and healthy tissue molecules give a considerable possibility for the targeted delivery of nanoparticles to tumor cells. To enable targeted deposition into a tumor mass, nanoparticles integrate therapeutic radionuclides that emit short-range radiation.

This includes radionuclides that emit alpha particles, low-energy beta particles, and Auger electrons. Heavy ions, neutrons, protons, and other forms of radiation have multiple applications in the treatment of various malignancies. These procedures presently concentrate on rays emanating from outside the body. External beams selectively irradiate the tumor mass while still administering a dosage to healthy tissue. In contrast to prior therapeutic methods, a possible technique would employ radiation-generating devices or antimatter implanted within the body to selectively irradiate a tumor. Through internal insertion, these devices emit different forms of radiation that specifically target the tumor mass. Internal radiation-generating or antimatter devices are now non-existent, and substantial advancements are necessary before they can be employed for the effective treatment of cancer. These characteristics include the choice of internal beam energy and radiation type. While internal radiation-generating devices remain theoretical, the necessary technology for their construction is advancing towards realization.

3.16. LASER

Laser-driven electron accelerators have been constructed. These accelerators are optical cavities that are ideally sized to the wavelength of light (Hemsing et al., 2014). Utilizing shorter-wavelength radiation would reduce the dimensions of these devices to the size anticipated for internal radiation-generating apparatus. The results indicate that precise target irradiation locations are attained by choosing suitable ion and energy combinations. The capacity to precisely target a region by selecting the ion and its energy is a beneficial characteristic that renders heavy ions a compelling option for external beam therapy applications and endorses their prospective utilization in an internal beam device. A ^{197}Au beam has a range of 0.20 to 1.91 cm for energies between 90 and 330 MeV/nucleon. Within this identical energy range, lighter ions yield higher ranges, whereas heavier ions result in shorter ranges. The internal device concept possesses the adaptability to irradiate the entire tumor mass by modifying the beam energy and radiation type. The absorbed dose constitutes merely a component of the therapeutic regimen. An effective therapeutic method directs the beam to the intended target volume. The beam should be checked and redirected as necessary. A heavy ion beam from an external source can be monitored with PET. If the particle range is significantly smaller than the tumor size, internal beam tracking becomes a secondary concern. An internal radiation-generating device would possess adequate beam control to preferentially deliver radiation to the tumor bulk. PDT is a non-ionizing radiation method that may serve as a feasible option for cancer therapy. It is more selective than alternative treatments and inflicts less damage on healthy tissue. It occurs selectively in the preferential absorption of photoreactive substances or photosensitizing agents by tumor cells. Enhancing the integration of photosensitization and conjugate agents is essential for the progression of the technology. Photodynamic therapy (PDT) presently administers the photosensitizing chemical to the patient, which preferentially accumulates in the tumor (Correia et al., 2021). A diode laser or LED emits light of the specified wavelength. The relevant wavelength is aimed at the tumor area, activating the photosensitizing drug while preserving healthy tissue. Upon activation, the photosensitizing agent converts a portion of its energy into molecular oxygen, generating excited oxygen that destroys cancer cells by oxidation. The efficient delivery of light to the tumor location remains a significant challenge. Preliminary PDT investigations focused on several melanoma forms due to the ability of near-infrared wavelengths to penetrate several millimeters through the skin to the tumor location. Recent advancements in light transport technologies enable access to deeper tumors. These delivery techniques broaden the spectrum of effective wavelengths and corresponding photosensitizing chemicals. Enhancing a photosensitization agent affinity for tumor sites is a problem, as the human immune system may target some compounds, hence diminishing the overall efficacy of the PDT method (Figure 3.17).

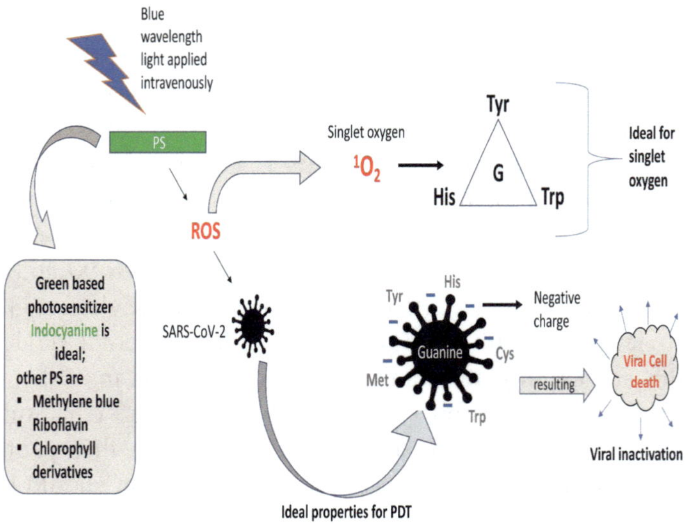

Figure 3.17. *Photodynamic therapy.*

Source: Tariq R, Khalid UA, Kanwal S, Adnan F, Qasim M. (2021). Photodynamic therapy: A rational approach toward COVID-19 management. J. Explor. Res. Pharmacol., 6(2):44-52. doi: 10.14218/JERP.2020.00036., Creative Commons License.

Current research endeavors involve employing lipoprotein shells and gold nanoparticle delivery technologies to improve deposition into tumors. Low-Coherence Interferometry (LCI) is a diagnostic imaging mode that integrates the benefits of Optical Coherence Tomography (OCT) with light scattering methodologies. Angle-resolved LCI employs this technique to differentiate scattering from subsurface tissue strata. Light scattering spectroscopy is employed to get structural information related to these tissue layers through the analysis of angular scattering data. This structural information distinguishes between healthy and sick tissue and elucidates their extent. OCT is the counterpart of ultrasound and constructs images through computer methods. The OCT approach analyzes a sample by merging backscattered photons with a reference beam through an interferometer. A light source with restricted coherence keeps the scattered photons out of phase with the reference beam upon collection at the detector. This enables the targeted measurement of scattered photons. Image contrast arises from variations in the refractive index within tissue. Besides the information obtained from variations in the refractive index, OCT can identify structures that modify the light within the imaged tissue. Polarization shifts signify alterations in collagen, the predominant protein in animals. Tissue exhibits translucence at red and near-infrared wavelengths. At wavelengths of 800–1300 nm, tissue penetration depths range from 1 to 3 mm. Despite being a valuable noninvasive imaging mode, OCT limited range has constrained its applicability. This approach is applicable to diagnostic studies of the pulmonary, circulatory, ocular, and skin. OCT has achieved the greatest success in ophthalmology, as the eye transparency allows for effective light penetration across it. OCT facilitates high-resolution analysis of the retina. OCT is utilized to assess plaque with enhanced clarity. Other uses include methodologies that integrate OCT with other diagnostic techniques, such as fluorescence and Raman spectroscopy. High-resolution retinal imaging facilitates the identification of the early start of glaucoma. Cornea imaging is enhanced as OCT can assess the eye topography and internal architecture. These are critical factors in enhancing refractive surgery outcomes. Significant accidents at major power reactors profoundly influence public perceptions and attitudes regarding technology that employ radiation or radioactive materials. Airport scanners employed for security assessments also affect public opinion. Similarly, the utilization of nuclear medicine techniques elevates public consciousness regarding the growing application of radioactive materials and their corresponding dosages. The radiation exposure experienced by aircrews and space tourists, along with the risk of solar flares, is a possible public health issue. Radionuclides that occur naturally include both cosmogenic and primordial types. Most cosmogenic radionuclides undergo decay via beta, gamma-ray, or X-ray emission; possess low to intermediate atomic numbers; and are generated through the interactions of cosmic radiation with target atoms in the atmosphere and Earth. The principal cosmogenic nuclides are ^{3}H, ^{7}Be, ^{14}C, and ^{22}Na. ^{14}C is the primary cosmogenic radionuclide responsible for internal exposure (Wieler, 2018). The majority of primordial radionuclides belong to the natural decay series of ^{232}Th, ^{235}U, and ^{238}U.The radiation emitted from the decay series of ^{238}U and ^{232}Th significantly contributes to the average effective dose received by the US population. The predominant primordial radionuclides are ^{40}K and ^{87}Rb. ^{40}K accounts for approximately one-third of both the external and internal effective doses originating from natural sources. Due to ^{87}Rb being a pure beta emitter, it provides negligible external dosage, and its contribution to internal exposure is lower than that of ^{40}K. The primary contributors to internal radiation exposure are isotopes of uranium, thorium, radon, polonium, bismuth, radium, and lead, originating from the decay series of ^{238}U, ^{232}Th, and ^{40}K.

3.17. RADON

Radon is generated from the decay series of uranium-238, while thoron is a component of the decay chain of thorium-232. Radon is classified as a distinct category in NCRP 160 because of its unique characteristics and substantial contribution to effective dose. The mean radon concentration in the United States is 43.3 Bq/m^3, while the average outdoor concentration is 15.1 Bq/m^3. The National Council on Radiation Protection and Measurements (NCRP) 160 adopts an effective dose conversion factor (DCF) of 10 mSv/WLM (working level month) for radon (Harley, 2020). The effective thoron dose conversion factor is between 3.3 and 3.8 mSv/WLM.

3.18. NUCLEAR WEAPONS

Human activities generate radionuclides and emit radioactive substances into the environment. These activities include atmospheric nuclear weapons testing, fuel cycle operations facilitating weapons production and nuclear power generation, radionuclides utilized in medicine, research, and industry, as well as nuclear reactor incidents. The radiation from weapons testing fallout has markedly diminished since the end of massive air testing around 1980. The radioactive material emitted after the Chernobyl disaster did not substantially affect radiation exposure in the United States. It was a substantial contributor throughout Europe and Asia. The majority of radioactive elements discharged from the Fukushima Daiichi incident were deposited in Japan; nevertheless, residual activity has been observed in adjacent regions. Protons and alpha particles, along with minor contributions from light nuclei, are the predominant kinds of space radiation. Four primary sources of space radiation impact the Earth and its atmosphere. The incident particulate radiation possesses adequate energy to initiate nuclear reactions and generate high-energy muons, electrons, photons, and neutrons. Neutrons are diminished by the environment, impart relatively low effective doses at sea level, and produce substantial doses at elevated altitudes. Solar-induced disruptions predominantly get noticed when they interfere with communication and electrical power networks. Solar storms are recognized for their ability to impair cellular phone services, global positioning systems (GPS), electrical power grids, and television and radio broadcasts. In March 1989, a solar storm caused disruptions to the electricity networks of the United States, Canada, and Sweden for several hours. The resultant damages and revenue loss were projected to be in the hundreds of millions of dollars. A solar flare in 1994 resulted in malfunctioning of communication satellites, disrupting television and radio services across Canada. The impairment of communication capabilities during a nuclear incident could substantially affect emergency management and associated operations. The Northeast blackout of 2003 was a significant power outage triggered by a solar event. This event impacted regions of the Northeastern and Midwestern United States, as well as Ontario, Canada. The blackout impacted around 10 million in Ontario and 45 million individuals across eight US states. The failure of power at a nuclear power plant has substantial consequences. In 2005, X-rays from a solar storm interfered with satellite-to-ground communications and GPS navigation signals for around 10 minutes (Singh et al., 2010). Solar flare radiation, or solar particle events (SPEs), are ejections of matter from the Sun. Their composition reflects the fundamental constituent properties of solar plasmas. The intensity of these SPEs poses a considerable radiation risk to astronauts in low Earth orbit (LEO), on lunar missions, and on forthcoming trips to other planets. Solar flare radiation is a crucial factor in the developing space tourism business, which may offer orbital and suborbital travel within the next decade. Electronic technology and their components are susceptible to SPEs. Cell phone communications, GPS, and radar broadcasts are susceptible to significant SPEs. The commercial satellites are vulnerable to significant occurrences such as the Carrington flare. Individuals operating in space would likewise face hazards. Astronauts and space tourists would have a restricted time frame from the initial detection of a significant solar particle event to seek refuge from energetic solar particles and photons.

CHAPTER 4

RADIATION DOSIMETRY

Contents

4.1. INTRODUCTION

In 1895, Wilhelm Roentgen (Figure 4.1) discovered X-rays, capable of penetrating materials (Martin and Harbison, 1980). Scientists have focused on the interaction between radiation and matter.

Figure 4.1. *Wilhelm roentgen.*

Source: https://upload.wikimedia.org/wikipedia/commons/3/3a/Wilhelm_R%C3%B6ntgen_1900_Portrait_%283x4_cropped%29.jpg, Creative Commons License.

Radiation encompasses energetic charged particles, including protons, alpha particles, electrons, and heavy ions; neutral particles, such as neutrons; and components of the electromagnetic spectrum, including X-rays and gamma rays. Radiation can modify the atomic and molecular composition of a medium by engaging with its atomic electrons and nuclei.

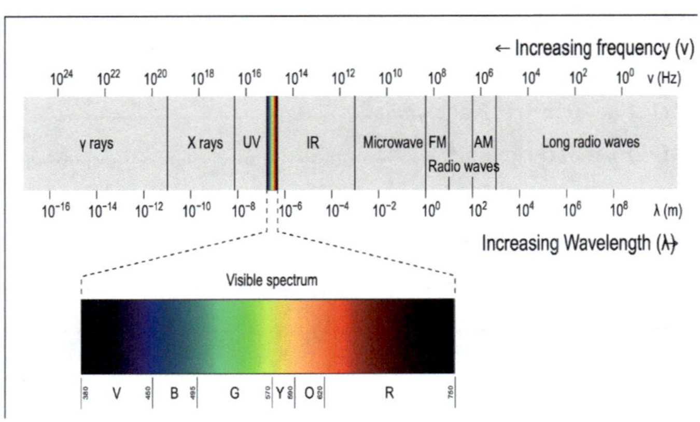

Figure 4.2. *Electromagnetic radiation.*

Source: https://upload.wikimedia.org/wikipedia/commons/1/14/EM_spectrum_updated.svg, Creative Commons License.

Radiation dosimetry includes the mechanisms of interactions and methods for measuring radiation and its consequences, relevant to disciplines such as biology and medicine. Radiation dosimetry involves the computation and assessment of energy received by an irradiated medium. When a projectile engages with a target, it is feasible to compute the energy dissipated by the projectile and transferred or transported to the target. According to wave theory, electromagnetic radiation engages with charged particles via forces applied by electric and magnetic fields. Due to the electric field's high magnitude compared to the magnetic field, interactions with charged particles mainly occur through the electric field.

4.2. ELECTROMAGNETIC RADIATION

Electromagnetic radiation (Figure 4.2) can occasionally release electrons when it interacts with matter. This phenomenon is referred to as the photoelectric effect. The wave theory of light postulates that the effect should manifest at any frequency of adequate intensity.

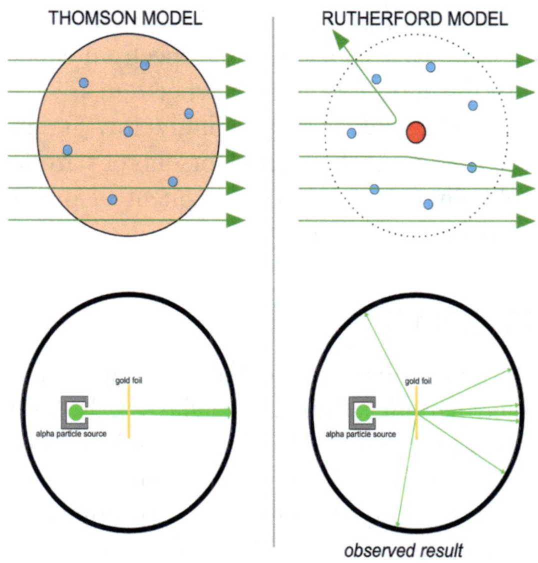

Figure 4.3. *Rutherford experiment.*

Source: https://upload.wikimedia.org/wikipedia/commons/f/f9/Geiger-Marsden_experiment_expectation_and_result.svg, Creative Commons License.

Experiments demonstrated that there exists a certain frequency beyond which the photoelectric effect does not occur, irrespective of light intensity. Einstein proposed that the energy of an electromagnetic wave is not constantly dispersed in space but rather sent in discrete packets, or quanta, known as photons, each possessing energy E = hv, where v represents the light frequency. Planck's constant, $h = 6.6262 \times 10^{-34}$ J s, was previously established through Planck's theory of blackbody radiation. The intensity of light is then ascertained by the quantity of photons. Einstein viewed photons as particles to elucidate the photoelectric effect. Einstein's elucidation of the photoelectric effect introduced a wave-particle duality to electromagnetic radiation. Certain interactions of electromagnetic radiation, including diffraction, are elucidated by wave theory. Other interactions, such as the photoelectric effect, can solely be explained by the particle theory, wherein photons engage with matter equivalent to particles, adhering to the principles of energy and momentum conservation. Electromagnetic waves may possess any frequency. Given that remains constant in a vacuum, high-frequency waves possess short wavelengths, whereas low-frequency waves have extended wavelengths. The electromagnetic radiation encompasses a broad or wider range of wavelengths and energy. Photons of waves with exceedingly small wavelengths possess similarly high frequency and energy. The and X rays possess sufficient energy to ionize atoms and disrupt molecular bonds. This segment of the spectrum is hence designated as ionizing radiation. At the opposite end of the spectrum are waves characterized by extensive wavelengths and correspondingly low frequencies, including microwaves and radio waves. The energy of photons from these waves are lower than atomic binding energies. This segment of the spectrum is hence designated as non-ionizing radiation.

4.3. EXPERIMENTS

By 1902, it was determined that alpha radiation emitted by uranium consisted of positively charged particles (Vallabhajosula, 2023). In 1911, Rutherford examined the findings of Geiger and Marsden studies with the scattering of alpha particles striking a thin gold foil to explain the nature of atoms. Several alpha particles altered their trajectory (Figure 4.3). Rutherford postulated that an atom comprises a nucleus that contains the positive charge and all of the mass. The lighter electrons he proposed are dispersed around this nucleus. When an alpha particle collides with an atom, its trajectory remains unaltered by the lighter electrons. However, the Coulomb attraction generated by the nuclear charge induces a change in trajectory (scattering). Rutherford computed the proportion of alpha particles expected to be deflected at specific angular directions. The experimental results corroborated this estimate, resulting in the acceptance of the Rutherford hypothesis. Bohr developed a planetary model for the dispersion of atomic electrons, wherein electrons orbit the nucleus. The

Coulombic attraction exerted on each electron by the nucleus maintains the electrons in their respective orbits. The most basic orbital shape, as suggested by Bohr, is round. As the electron accelerates, it is expected, per classical electromagnetic theory, to emit energy. The electron would rapidly dissipate all its energy, resulting in the collapse of the atom. A stable atom, as postulated by Rutherford, would be unfeasible. The quantum theory was established by Erwin Schrödinger in 1925. De Broglie's achievement in deriving quantized atomic orbits by conceptualizing an electron as a wave initiated the pursuit of a wave theory for particles. Atoms in excited states rapidly revert to the ground state, emitting energy often as light. When the electron of hydrogen is excited to the L-shell, it transitions back to the K-shell. The electron becomes more constrained and loses energy equivalent to the difference in shell binding energies of 10.2 eV, which is emitted as a photon. This ultraviolet spectral line is bifurcated into two closely spaced lines in terms of energy. This phenomenon is referred to as fine structure. An orbiting electron forms an electric current, which generates a magnetic field. This orbital magnetic field interacts with other magnetic fields, modifying the energy of shells. The nucleus includes all the positive charge of an atom; it became evident that an unexplored force must exist. The repulsive Coulomb force between protons would render nuclei unstable, preventing the existence of the Rutherford nucleus. To mitigate this repulsion, an attractive force must be exerted between protons. This nuclear force is between all nucleons (protons and neutrons).

4.4. SCATTERING

Scattering experiments can be utilized to measure the range of the nuclear force and the dimensions of a nucleus. In the analysis of 5.5 MeV alpha-particle scattering by gold conducted by Rutherford, the Coulomb force accurately forecasts the scattering pattern (Podgoršak, 2016). It follows that alpha particles did not interact with the nucleus; otherwise, the scattering would be influenced by the nuclear force. Direct backscatter occurs with a collision. As the alpha particle approaches the nucleus, it decelerates due to the repulsive force. At a certain point, characterized by the smallest separation, the kinetic energy is zero, and the total energy equates to the potential energy, expressed as $kZ_1Z_2e^2/r$, where Z_1e and Z_2e represent the charges of the alpha particle and the nucleus, respectively. By equating this to the original 5.5 MeV and assigning $Z_1 = 2$ and $Z_2 = 79$ for the alpha particle and gold, the minimum separation is determined to be 4×10^{-12} cm. The absence of contact results in an overestimation of the nuclear dimensions and the extent of the nuclear force. The interaction of radiation with matter is a stochastic process, meaning it is random. It is impossible to ascertain which particle will engage with which atom or the number of particles that will interact in a beam of particles. Similar atoms, molecules, or cells exposed to the same beam may undergo varying quantities and kinds of interactions. Such processes can only be characterized by probabilities derived from several collisions. Upon heating a metal, its atoms collide with adequate energy to liberate certain loosely bound outer-shell electrons. This phenomenon is referred to as thermionic emission. The filament, typically composed of tungsten in a wire coil configuration, is heated by a low voltage power source.

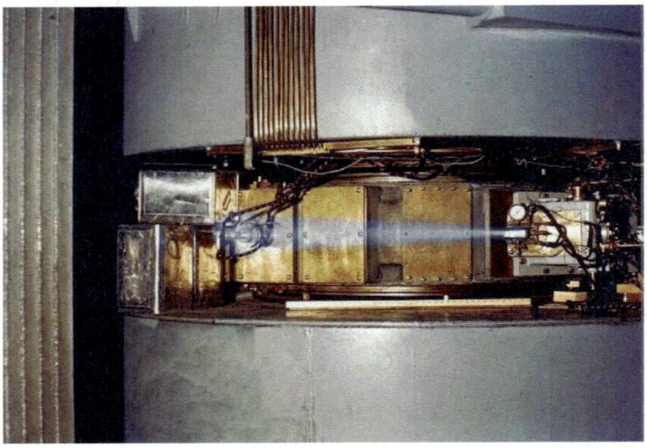

Figure 4.4. *Cyclotron.*

Source: https://upload.wikimedia.org/wikipedia/commons/b/ba/Cyclotron_with_glowing_beam.jpg, Creative Commons License.

4.5. PROTONS

Proton beams primarily originate from hydrogen gas. Electrons emitted from an electron cannon cross the atmosphere and ionize hydrogen atoms. Electric fields displace the light electrons, allowing the heavier protons to be accelerated across a potential difference, as demonstrated in an electron gun, resulting in a low-energy proton beam. The identical process can be employed to generate beams of positive ions from heavier nuclei. These devices are referred to as proton or ion guns. A beam produced by such a gun can be introduced into an accelerator, such as a cyclotron, (Figure 4.4) for additional acceleration to elevated energy levels (Livingston, 1952). The mechanisms generating particles (or photons) are stochastic, leading to fluctuations in the quantity of beam particles with respect to position (space) and time. A beam is characterized by its mean particle count. Charged particles generally interact via the spherically symmetrical Coulomb force. Nuclear charged particles, including protons and alpha particles, can interact with other nuclear particles through the significantly stronger yet short-ranged nuclear force. Classical electromagnetic theory demonstrates that accelerating charged particles emit electromagnetic radiation. Any interaction among charged particles is expected to generate electromagnetic radiation. The quantum theory indicates that non-radiative interactions, such as elastic scattering, are more probable. In elastic scattering, the kinetic energy of the system remains conserved. For a target nucleus at rest, the initial kinetic energy is the same to that of the projectile. Subsequent to the collision, this energy is distributed between the scattered projectile and the recoiling nucleus. If the nucleus possesses greater mass than the projectile, only a minimal portion of energy is imparted to it, allowing the projectile to largely retain its initial energy. Electrons are not affected by the nuclear force. The elastic scattering of electrons by nuclei occurs solely through the Coulomb force. Elastic scattering of charged nuclear particles by nuclei predominantly happens via the Coulomb force. In proximity to a nucleus, the nuclear force facilitates the scattering process.

4.6. RADIATIVE SCATTERING

Inelastic radiative scattering occurs when a charged particle loses energy to a photon of electromagnetic radiation via interactions with nuclei, a phenomenon known as bremsstrahlung radiation (Apte and Bhide, 2024). In inelastic non-radiative scattering, energetic particles can induce nuclear excitations and nuclear reactions. For particles capable of experiencing the nuclear force, these interactions occur during close nuclear collisions and hence possess a significantly lower probability than elastic scattering. A proximate interaction is required for electrons, as they do not engage with the nuclear force. These interactions are insignificant for electrons. Advanced theories expand upon Rutherford's model by incorporating relativistic particle velocities, particle spin, and the ability of energetic nuclear particles to approach a nucleus closely enough to experience both the nuclear and Coulomb forces. In many cases, these do not influence the magnitude of the Rutherford cross sections. The Rutherford cross sections will be employed in this analysis of elastic nuclear scattering projects. A beam of particles crossing matter may interact with nuclei and atomic electrons, resulting in the scattering of some particles beyond the beam's geometric limits. This lateral dispersion results in a decrease in beam fluence. When a heavy particle, such as a proton, interacts with an electron, it is deflected at an exceedingly small angle, with a maximum scattering angle of approximately 0.03°. Interactions with nuclei mostly involve small-angle scattering. As the particle moves through a medium, multiple interactions with nuclei lead to lateral dispersion. Electrons undergo greater large-angle scattering from electrons and nuclei compared to heavier particles. The frequent interactions of electrons as they go through a medium can lead to considerable lateral dispersion of the beam, which intensifies with depth. Electron beams disperse significantly more than heavy charged-particle beams. A charged particle, when moving across a medium, interacts with and dissipates energy to atomic electrons and nuclei. The long-range Coulomb force leads to extensive and frequent contacts, causing a loss of energy. There are two primary categories of interactions involving charged particles. In radiative interactions, each charged particle that interacts with another charged particle undergoes acceleration and, in accordance with classical electromagnetic theory, emits electromagnetic energy known as bremsstrahlung radiation. This is typically the dominant kind of radiative energy dissipation. In the collision interactions, the charged particles dissipate energy to atomic electrons and nuclei during collision events. The collision energy lost to a nucleus is insignificant when compared to the collision energy lost to an electron for electrons and positrons. Heavily charged particles can dissipate more energy to nuclei upon a direct impact than to electrons. Cross sections for inelastic scattering by atomic electrons must be computed using quantum theory with suitable wave functions to mathematically describe atomic energy levels (Burgess and Percival, 1968). Prior to the understanding of atomic states, an estimate was derived from Thomson. He derived an expression connecting the energy dissipated by a charged projectile from classical elastic scattering. He determined that for rapid projectiles with velocities significantly exceeding that of the electron orbital velocity, the interaction period is short for the electron to be regarded as free and stationary. The interaction period is so short that the collision force surpasses the binding force between the electron and the nucleus, rendering the electron effectively stationary and free. Rapidly moving charged particles interacting with an electron typically undergo minor scattering angles and dissipate a small fraction of their energy to the electron. This is especially applicable to heavily charged particles. The dispersion can be approximated by considering the projectile as moving in a linear or straight trajectory with minute energy loss. The straight or linear trajectory assumption is valid, particularly for heavy charged particles, as they undergo minimal scattering angles.

4.7. SOFT COLLISIONS

Soft collisions are inelastic and lead to atomic excitations or ionizations, with the electron receiving minimal or no kinetic energy. Soft impacts and their minimal energy losses predominate due to their occurrence at substantial impact parameters, which possess correspondingly large cross sections. A medium with a high atomic number is often dense, leading to a subsequent reduction in the stopping power for electrons and positrons. Lead possesses approximately 20% less electrons per gram than aluminum, and the ionization energy of lead is almost six times that of aluminum (Prasad et al., 1998). The electronic mass collision stopping power in lead is approximately 40% lower than in aluminum for 100 MeV protons and around 30% lower for 10 MeV electrons. A substantial amount of radiation is generated in the vicinity of a high Z target. Interactions with a nucleus generate more bremsstrahlung than interactions with an electron. The bremsstrahlung generated by the Z electrons of an atom is significant. For light charged particles, an additional potential mechanism for radiative energy loss is Cherenkov radiation, which occurs when the particle velocity in a medium surpasses the phase velocity of light within that medium. The radiative energy loss due to Cherenkov radiation occurs in the blue to ultraviolet spectrum and is sufficiently less relative to bremsstrahlung to typically be ignored. A positron serves as the antiparticle of an electron. It possesses an identical mass and rest energy to that of an electron, yet carries a positive charge that is equal in magnitude to the electron charge. When matter encounters antimatter, they annihilate, resulting in the generation of photons. The main positron-electron annihilation emerges when the positron has expended nearly all its energy, with the rest mass energy of the positron transformed into photon energy. Significantly fewer annihilations occur while the positron is in motion. During these in-flight annihilations, the kinetic energy of the positron is also converted into photons. Radiative energy loss due to in-flight annihilation is often minimal, however not insignificant when compared with bremsstrahlung radiation loss. Heavy charged particles exhibit minimal deviation when interacting with electrons, and the majority of their scattering interactions with nuclei occur at minimum angles. Due to their linear trajectories, there is minimal scattering into or out of the CA. Heavy charged particles, such as protons, can induce nuclear reactions. Due to the necessity of a proximate interaction with a nucleus, their cross sections are significantly less than those of Coulomb scattering. Nuclear processes, however, diminish beam fluence as particles are absorbed by nuclei to create compound nuclei. At minimal velocities, a particle remains in proximity to atoms long enough to acquire atomic electrons. This substantially diminishes fluence. If a proton takes an electron, it ceases to be a beam particle. In the absence of nuclear reactions, the particle counts and fluence of a heavy charged particle beam would remain almost constant until approaching a depth near the range. Due to nuclear processes, both the quantity and fluence gradually diminish with depth. Electrons are categorized as primary or secondary electrons. Primary electrons in electron beams refer to those that are part of the incident beam. For all other beams, primary electrons are those activated by beam particles. First generation secondary electrons are those initiated by primary electrons; second generation electrons are those initiated by first generation electrons; and so forth (Hachenberg and Brauer, 1959). All generations will be designated as secondary electrons. Confusion arises regarding the designation of primary and secondary electrons when a primary electron imparts energy to a secondary electron. This is due to the fact that the two electrons are indistinguishable particles. The electron with lesser energy is designated as the secondary. Stopping power allows for the calculation of the mean energy dissipated by a particle during collisions with atomic electrons, as well as, for electrons and positrons, the mean energy lost to bremsstrahlung radiation in radiative interactions. The photons of bremsstrahlung radiation possess an extended range. Only collision interactions with electrons result in absorbed energy. Another interaction for heavy charged particles is nuclear processes. These will be ignored because to their significantly smaller cross sections compared to the Coulomb interaction cross sections. Collision interactions lead to atomic excitations or ionizations and are classified as inelastic due to the energy expended in overcoming shell binding energies or excitation energies. Inner-shell electrons necessitate more intense collisions for excitation or ionization. Hard collisions occur less frequently. The majority of interactions involve outer shell electrons. Following excitation or ionization, a shell vacancy is occupied by an electron from a higher energy shell. During this process, the atom release energy equivalent to the energy it absorbed, which is released as radiation.

4.8. ELECTROMAGNETIC RADIATION (EMR)

EMR interacts with atomic electrons and nuclei upon entering matter. EMR was regarded as a wave, with interactions arising from its electric and magnetic fields. Certain interactions can be explained by this wave model. Due to the absence of charge and mass in photons, their interaction cross sections, or probability of interaction, are smaller than those of charged particles. Photons can lose energy through absorption or scattering when they contact with atoms. Energy dissipated to electrons leads to atomic excitations or ionizations. An electron released during an ionization event is propelled with a certain amount of kinetic energy. This electron will undergo multiple interactions during its trajectory, resulting in various ionizations and excitations, while dissipating energy as characterized by the stopping power (medium). Photons function as a catalyst. They impart energy to electrons, which then release that energy into the medium. The photons are classified as indirectly ionizing radiation, while charged particles are categorized as directly ionizing radiation. Following the excitation or ionization event, an atom possesses surplus energy and a vacancy in the shell from where the excited or ionized electron originated. Within around 10 to 8 seconds, the vacancy is occupied by an electron transitioning from a higher energy shell to the lower energy vacancy (Li et al., 2017). During this process, the atom expends energy equivalent to the difference in the binding energies of the two shells. The lost energy either generates a photon of characteristic radiation from the atom or is transferred to another orbital electron, resulting in its ejection or release as an Auger electron. Vacancies in atomic shells with low binding energy are often filled by the emission of Auger electrons, whereas vacancies in shells with high binding energy are generally filled by the emission of characteristic radiation. Low atomic number elements mainly occupy shells by Auger electron emission. High atomic number elements predominantly occupy K-shell vacancies, resulting in the emission of characteristic radiation. L-shell vacancies are predominantly filled through Auger electron emissions, while vacancies in shells above L are primarily filled via Auger electron emission. The classical theory of electromagnetic radiation scattering by charged particles is founded on electromagnetic theory, which conceptualizes radiation as a wave. When an electromagnetic wave strikes a charged particle, the oscillating electric field of the wave induces oscillatory motion in the particle at the same frequency as the wave. The charge emits a wave, referred to as the scattered wave, which possesses the same frequency. In wave theory, there is no alteration in frequency or wavelength due to scattering. The electric field of the wave can induce oscillation in atomic electrons and nuclei.

4.9. COMPTON SCATTERING

Compton's investigations demonstrated that the wavelength of scattered electromagnetic radiation exceeds that of the incident wavelength (Jain and Tzoar, 1978). Compton determined that, in this case, electromagnetic radiation behaves as though it is composed of particles (photons), with scattering governed by the principles of energy and momentum conservation. He considered the scattering of a photon by a free electron. In Compton scattering, atomic electrons operate independently of one another. Each electron may undergo a Compton interaction with a photon from the irradiating beam, and there is no correlation among the scattered photons. The dispersed photons are hence incoherent, and Compton scattering is also referred to as incoherent scattering. Rayleigh scattering is an interaction involving a photon and electrons that are associated with an atom before and after the interaction. The conventional explanation postulates that the electric field of an incoming electromagnetic wave induces vibrational motion in atomic electrons. The accelerated electrons produce electromagnetic radiation, referred to as the scattered wave, at an angle relative to the incident direction. All electrons within an atom oscillate coherently, resulting in radiated waves that are in phase and correspond to the incident frequency. The amplitudes of the radiated waves vary due to the differing binding energies of the electrons. The resultant scattered electromagnetic wave is a coherent aggregation of varying amplitudes but identical frequencies and phases. Rayleigh scattering is hence termed coherent scattering. Following the collision, the electrons return to their designated atomic levels without any energy being imparted to them. As the atom is neither excited nor ionized, this interaction is more likely to occur at low photon energy, where binding energies limit Compton scattering. Photoelectric interactions, similar to Rayleigh scattering, emerge at photon energies where atomic electron binding energies are considerable. A photon is absorbed by an atom, resulting in the release of an electron from a bonded atomic state. This can only occur when hv exceeds or lowers the binding energy of the state from which the electron is released. The released electron is referred to as a photoelectron. A free electron cannot absorb a photon, as this would violate the simultaneous conservation of energy and momentum. The photoelectric effect necessitates the involvement of electrons linked to an atom that recoils to maintain the initial zero momentum. Given that the electron cannot exist freely, it is anticipated that the chance of interaction is higher for the bound inner-shell electrons compared to the outer-shell electrons, which are close to a free state. Pair production is an absorption interaction wherein a photon disappears, resulting in the formation of an electron-positron pair (Baur et al., 2007). Pair production cannot take place in a vacuum. It must occur in the presence of an additional particle that recoils post-interaction to ensure momentum conservation. The interaction occurs between the electromagnetic field of the photon and the Coulomb field of a charged particle, which may be a nucleus or an electron. Interactions between photons and nuclei are referred to as photonuclear reactions. During these interactions, the input photon is absorbed by a nucleus, which subsequently produces a proton, neutron, or fission fragments. The most prevalent of these, the emission of a single neutron. Photonuclear reactions can only take place if the photon energy surpasses the binding energy of a nucleon within the target nucleus. Excluding deuterium (2H) and beryllium-9 (^{9}Be), the typical binding energy of a nucleon is around 8 MeV, which serves as the threshold for photo nuclear processes. The cross sections for photonuclear reactions are significantly less than those for photon interactions with atomic electrons, rendering them negligible. In pair production involving a nucleus, the original photon energy is distributed among three particles: the electron, the positron, and the nucleus. The threshold represents the minimum photon energy required for the interaction to take place.

4.10. EMR INTERACTIONS

The interaction of electromagnetic radiation with matter is fundamentally attributed to the force applied on charges by the electric field of the wave. The dense nucleus of an atom exhibits significantly less reaction to the field compared to an electron. The scattering of the wave from an atom is mostly attributable to interactions with atomic electrons. The Thomson cross section is invariant with respect to the frequency and energy of electromagnetic radiation and forecasts identical backward and forward scattering. These align with experimental results for low photon energy of hv less than approximately 10 keV. At elevated energy, the dispersed radiation, as discovered by Compton, has a lower frequency than the incident radiation, and the Thomson cross section is large. In Thomson scattering, no energy is imparted to the electron, in contrast to Compton scattering. For low-energy photons, Compton equations indicate that the energy transferred is minimal. The scattering cross section for photons is significantly smaller than that for charged-particle interactions. The photons exhibit less interactions compared to charged particles. At low photon energies, where atomic binding energies are significant, two interactions become competitive with Compton scattering. These phenomena are Rayleigh scattering and photoelectric absorption. The interaction cross sections for both interactions, involving bound atomic electrons, should be calculated using quantum theory and suitable atomic state wave functions. In Rayleigh scattering, atoms remain unexcited and un-ionized, resulting in no energy transfer to electrons. The dispersed wave contributes to the spectrum of photons within a material. In the photoelectric interaction, an incident photon is absorbed by an atom, resulting in the ejection of an atomic electron, referred to as the photoelectron. The interaction cross section can alone be derived from quantum theory. At photon energies exceeding the K-shell binding energy, atomic electrons are free, resulting in a photoelectric cross section that is considerably less than the Compton cross section. The cross section rises with diminishing photon energy, reaching a maximum known as the K-edge, when the photon energy is equivalent to BE_K. At photon energies just below the K-edge, the cross section declines due to a reduced number of atomic electrons available for interaction.

4.11. PHOTOELECTRIC EFFECT

Photoelectric effects predominate over Rayleigh scattering at low photon energy, where both interactions are substantial (Cesareo et al., 1992). Pair formation is the sole cross section that escalates with photon energy, commencing at an energy of 1.022 MeV. Compton scattering occurs across all photon energy levels. At low energies, when electrons cannot be regarded as free, Compton scattering diminishes while Rayleigh scattering and the photoelectric effect increase. At low photon energies around shell binding energies, the photoelectric effect predominates. There exists a spectrum of frequencies for which Compton scattering is the interaction, even at the initial energies of pair production. This range is highly based on the atomic number (Z) of the element.

A photonuclear reaction is an absorption process wherein a nucleus absorbs a photon, resulting in the emission of a proton, gamma ray, fission fragments, or one or more neutrons. Neutrons emitted from these materials integrate into the beam, resulting in neutron contamination. In radiation therapy, these neutrons administer an extra dose to the patient by mobilizing protons. A charged particle encountering another charged particle undergoes acceleration, due to the Coulomb interaction and emits electromagnetic energy known as bremsstrahlung radiation (Figure 4.5).

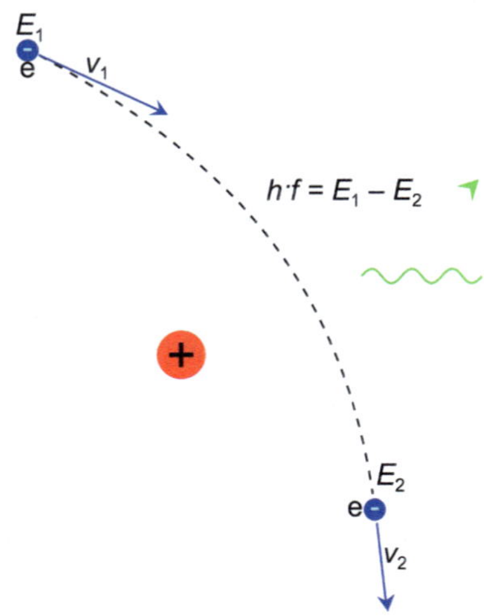

Figure 4.5. *Bremsstrahlung radiation.*

Source: https://upload.wikimedia.org/wikipedia/commons/1/1e/Bremsstrahlung.svg, Creative Commons License.

Some elements are crucial for the generation of bremsstrahlung X-ray beams. A metal component, known as the filament, is heated to a sufficiently high temperature to release the loosely bound electrons. A method or method for propelling electrons from the filament to the target. This can be accomplished by delivering a high voltage, often in kilovolts, between the filament and the target. A target material for electron impact is crucial. The target must possess a high atomic number to enhance bremsstrahlung interactions. The radiation yield of 100 keV electrons in tungsten is around 1%. The frequent collisions of incident electrons transfer energy to the atomic electrons of the target, which ultimately manifests as heat. The target must possess a high melting point. The target employed is tungsten, characterized by an atomic number of 72 and a melting point of 3,370°C. The electrons with minimal impact parameters experience significant energy loss and generate high-energy photons, whereas electrons with substantial impact parameters undergo minimal energy loss and emit low-energy photons. Kilovoltage X-ray beams, widely utilized in diagnostic imaging, are generated in X-ray tubes. Electrons commence from a state of rest at the filament and impact the target with a kinetic energy equivalent to the accelerating voltage, kV. The resultant photon beam is designated based on the accelerating voltage. With an accelerating voltage of 150 kV, electrons impact the target with an energy of 150 keV, generating a 150 kV X-ray beam with a maximum photon energy of 150 keV. The tube components are contained within an enclosure. The vacuum is essential to prevent spontaneous combustion of air that may arise from the generated heat. In kV beams, the bremsstrahlung generated in the target originates primarily from the surface, as X-rays produced within the interior are significantly reduced by photoelectric absorption prior to their escape from the target. In diagnostic imaging, the patient and detector are positioned within the beam trajectory.

4.12. BEAM HARDENING

The beam exiting the target crosses the tube housing and filters, each of which preferentially attenuates lower photon energy. Upon leaving the housing, the beam traverses optional external filters to further diminish low energies. The elimination of low-energy photons, known as beam hardening, enhances the average energy of the beam and its penetration capacity. In imaging, low-energy photons fail to enter the patient and do not provide information to the detector. The patient would be subjected to excess radiation exposure. Bremsstrahlung radiation generated by MeV electrons primarily occurs in the forward direction, consistent with electromagnetic theory (Korol and Solov'yov, 2014). An electron interacting with a nucleus undergoes acceleration along the axis connecting it to the nucleus. This acceleration can be divided into a component parallel to the velocity and a component perpendicular to the velocity. Electromagnetic theory postulates that each of these components preferentially emits radiation in the forward direction for high-energy electrons. A patient undergoing radiation therapy is positioned in alignment with the beam. A flattening filter is typically employed to produce a uniform beam directed at the patient. The filter further diminishes the beam intensity. Adjustable collimators determine the dimensions of the beam. Electrons emitted from photon interactions with the flattening filter and collimators are incorporated into the resultant X-ray beam. MV X-ray beams are contaminated with electrons. Beams above 10 MV contain photons with energy of 0.8 MeV. These photons can engage in photonuclear interactions with the flattening filter and collimators, thereby adding neutrons into the beam. X-ray beams of 10 MV or higher are also contaminated by neutrons. Flattening filter-free beams are generated by eliminating the filter. The beam directed towards the patient is the forward peaked beam originating from the target. One advantage of these beams is their reduced electron contamination, resulting in a lower skin dose. A secondary advantage is an increased dose rate due to the absence of beam attenuation by the filter. The beam configuration is suitable for irradiating a limited cross-sectional area in a patient. The beam may be traversed over a patient to irradiate an expanded cross-sectional region. The assessment of dose to a medium exposed to a low-energy X-ray beam relies on the measurement of air collision Kerma of the beam, then converted to collision Kerma for the medium (Omondi, 2021). The initial step is to quantify the Kerma of the beam-air interaction. Air ionization chambers are the main detectors employed for this purpose. Regardless of whether the irradiating beam consists of charged particles, photons, or neutrons, it is the charged particles that administer the dose. A charged particle moving through a gas experiences collision interactions, including ionization and excitation, as well as bremsstrahlung interactions. The majority of ionizations result from collision interactions; however, some are generated by bremsstrahlung radiation as it exits the gas. Each ionization generates an ion pair, consisting of a positively charged ion and a free electron.

4.13. CHARGES

Charges within the chamber are propelled by the electric field and decelerated by collisions. This yields an average or drift velocity that is directly proportional to the electric field. The slow movement of ions allows some negatively charged oxygen ions to discharge their charge to positively charged ions prior to collection. This phenomenon is referred to as ion recombination. The outcome is that the collected charge is less than the generated charge. This is quantified as ion collection efficiency. Elevated exposure rates and diminished chamber voltages yield reduced collection efficiency. At elevated voltages, ion recombination diminishes due to the accelerated drift of ions, which reduces their chance for recombination. Cavity theory correlates the dosage delivered to the cavity with the dose administered to the surrounding medium. The dose to the cavity from photon beams may result from electrons entering the cavity from the surrounding medium and from photon interactions occurring within the cavity. The dose to the cavity from neutron beams may result from charged particles entering the cavity from the surrounding medium, as well as from neutron interactions occurring within the cavity itself. The dose to the cavity from electron beams results from primary and secondary electrons entering the cavity from the surrounding medium. The dose to the cavity from heavy particle beams may result from heavy particles entering the cavity as well as electrons from the surrounding medium entering the cavity. The gas volume is minimal and devoid of CPE. Bragg and Gray aimed to ascertain the dose delivered to a medium, such as water, exposed to gamma radiation. An ionization chamber with a specified volume was positioned within the medium. The dose (air) was quantified from the accumulated charge. Gray devised a technique for translating this dose to the dose applicable to the continuous medium, specifically the medium devoid of the cavity (Enger et al., 2010). The initial boundary condition stipulates that the cavity dimensions are significantly smaller than the range of the incoming particles. The dose to the cavity is attributable exclusively to charged particles traversing it. The energy dissipated in the cavity can be determined from the stopping power. This criterion necessitates that the dosage to the cavity from photon and neutron interactions be minimal. The BG requirement stipulates that the cavity is sufficiently diminutive to avoid disrupting the directional and velocity distribution of the particles traversing it. The cavity must possess identical scattering properties to those of the medium. When this condition is met, the particle fluence within the cavity is equivalent to that in the continuous medium. Gray recognized that the cavity, differing from the medium, scatters particles in a distinct manner. At any location within the continuous medium, there are forward-moving particles and backward-moving particles that have been backscattered from the region beyond that point. If the cavity has a higher atomic number (Z) than the surrounding medium, the forward fluence of particles within the cavity is diminished compared to that in the continuous medium, since a greater number of particles are backscattered at the high Z interface than in the continuous medium. The alteration in forward fluence within the cavity also influences backscattering at the downstream terminus of the cavity. Due to varying scattering, particles arriving at the downstream end of the cavity possess different energy and trajectories compared to those in a continuous medium. This also influences backscattered fluence. The fluence in the cavity differs from that in the continuous medium. For MV photons, the majority of electrons set in motion possess energy significantly above 17 keV. The exposure to the air is mostly attributable to crossing electrons. A cavity can be regarded as a Bragg-Gray (BG) medium under photon irradiation for charged particles if its atomic number (Z) is comparable to that of the surrounding medium and is significantly less than the range of the predominant electrons in motion. Neutrons are categorized based on their kinetic energy. Similar to all other forms of radiation, neutrons can interact with atomic electrons and nuclei upon entering matter. Due to the electrical neutrality of neutrons, there is an absence of long-range Coulomb interaction. Similar to photons, neutrons exhibit significantly less interactions over the same distance compared to charged particles. The minimal point size of an electron renders its interaction cross sections small in comparison to those with nuclei, including protons. Only interactions with nuclei hold significance. Similar to photons, neutrons transport energy to charged particles, which subsequently transfer that energy to the medium predominantly via interactions with atomic electrons, as elucidated by the medium stopping power. The charged particles in motion are protons and other nuclei. Neutrons constitute indirectly ionizing radiation, and the energy they impart to charged particles is quantified as Kerma. Neutrons engage with nuclei in many ways, based on neutron energy and the specific nucleus involved. Interactions with a certain nucleus may significantly differ from those with the nucleus of one of its isotopes. The primary forms of interactions of interest are scattering and absorption. This is the non-absorption interaction. The neutron does not infiltrate the nucleus, and its incident kinetic energy is conserved, being distributed between the scattered neutron and the recoiling nucleus. The interaction, mediated by the nuclear potential or nuclear force, necessitates a direct impact and can be regarded as hard sphere scattering. An estimation of the cross section can be conducted for high-energy neutrons. A 20 MeV neutron is precisely confined with a de Broglie wavelength of 6×10^{-13} cm (Pynn, 2009). The cross section is subsequently estimated using the geometric area defined by the combined radii of the neutron and nucleus, specifically, 10^{-25}cm^2. The cross section increases at decreasing energy as the neutron becomes more delocalized. The interaction cross sections for neutrons are similar to those of photons.

4.14. RADIATION CAPTURE

Radiative and non-radiative capture can occur at any neutron energy; however, they are more probable with slow neutrons. The capture cross section for thermal neutrons may exceed the geometrical cross section, as the de Broglie wavelength of a thermal neutron is approximately 1.8×10^{-8} cm, which is 10^5 times larger than a nuclear radius. A neutron is distributed over a wide distance, resulting in a significantly larger cross section compared to a localized or concentrated particle. The capture cross section inversely correlates with neutron velocity. The capture of slow neutrons can be understood as proportional to the duration a neutron residence time near a nucleus. Neutron beams are invariably contaminated with photons due to inelastic scattering and radiative capture. No detector exists that only responds to photons or neutrons. A single detector is incapable of measuring dosage in neutron beams, as its response cannot be translated into dose for the medium due to the differing conversion factors for neutrons and photons. Dosage can, however, be quantified with two paired detectors with established photon and neutron calibrations. The primary benefit of solid detectors compared to gas ionization chambers is their higher density, allowing for a more compact design while maintaining equivalent signal output. Solid detectors possess a crystalline structure. A pure crystal does not exist. The deviation from purity is termed defects. Point defects arise within the crystal lattice. These may be dislocation defects characterized by the absence of an atom at an occupied crystal site or the incorporation of additional atoms into the crystal that occupy positions where an atom is not ordinarily present. The latter are referred to as interstitial atoms. Atoms distinct from those of the crystal may be integrated into the crystal structure. If this emerges organically during the growth phase, these are referred to as contaminants or impurities. Dopants, or foreign atoms, may be deliberately introduced throughout the growth process to modify the crystal reaction. Dopant atoms occupy sites in the crystal lattice held by a silicon atom. A group V dopant atom, like phosphorus, possesses five valence electrons. When it occupies a lattice site, it shares four of these with adjacent Si atoms. The additional electron is loosely bonded and virtually free to traverse the crystal, rendering it an electron donor or n-type silicon. It is important to note that n-type silicon is electrically neutral yet possesses an excess of electrons relative to pure silicon. A group III dopant atom, like boron, possesses three valence electrons. When situated in a lattice location, it may establish covalent bonds with only three of its silicon neighbors. The hole together with the resultant crystal, functions as an electron acceptor or p-type silicon. An electron from an alternative location may occupy this vacancy. Thus, the holes, akin to the electrons in n-type silicon, are fundamentally unbound or free. It is important to note that p-type silicon is electrically neutral yet exhibits a shortage of electrons (holes) relative to pure silicon.

4.15. RADIOGRAPHIC FILM

Radiographic film (Figure 4.6) comprises a thin coating of radiation-sensitive emulsion applied to one or both sides of a transparent flexible substrate (Clough, 2001). The emulsion comprises gelatin infused with silver halide crystals, called silver bromide (AgBr). The crystal is a cubic lattice constituted by ionic bonds between Ag^+ and Br^-. The size of a typical crystal is 1 gram, and there are around 10 billion crystals per cubic centimeter. During the crystal growth process, certain interstitial Ag^+ ions are integrated into the crystal lattice. A minimal quantity of sulfur introduced to the gelatin interacts with silver halide to produce silver sulfide (Ag_2S), which becomes integrated into a crystal. Both silver and sulfur exhibit significant electronegativity. The defect, termed the sensitivity speck, along with the interstitial Ag^+, is crucial to the film reaction. When film is exposed to light or ionizing radiation, electrons from the bromide ions are released and adhere to the crystal sensitivity speck. The neutral bromine atom disperses into the gelatin. The negatively charged sensitivity particle attracts interstitial Ag^+ ions, which are neutralized to form metallic Ag. The silver deposits are referred to as the latent image.

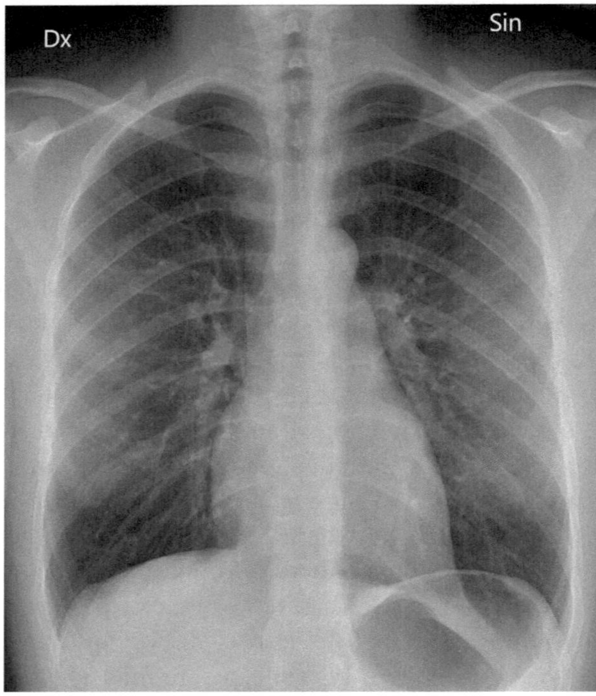

Figure 4.6. *Radiographic film.*

Source: https://upload.wikimedia.org/wikipedia/commons/a/a1/Normal_posteroanterior_%28PA%29_chest_radiograph_%28X-ray%29.jpg, Creative Commons License.

A crystal with a latent image is transformed into a grain of silver during the film development process. Developing stage is a reduction process in which electrons are introduced by the developing agent. Some electrons migrate to the sensitivity speck, which attracts additional silver ions and neutralizes them. Crystals subjected to radiation have lost their bromide ions, and the remaining Ag^+ ions are transformed to silver. Crystals possessing a latent image are converted to metallic silver. Crystals devoid of a latent image are unaltered. In the fixing stage, the fixing agent interacts with the unaltered silver halide crystals to produce a water-soluble complex. Washing, a final stage, the fixing agents and unaltered silver halide crystals are removed from the film, leaving the silver grains. Upon processing, regions of a film subjected to higher radiation levels exhibit an increased density of black metallic silver grains, resulting in a discernible image in varying shades of gray. Radiochromic film comprises a delicate layer of a material infused with a radiosensitive dye of microcrystalline monomers, situated on a transparent substrate. Due to the elevated spatial resolution, optical density is typically measured using a significantly smaller light source than that employed for radiography film, as utilized in a laser scanner or microdensitometer. Thermoluminescence, the release of light upon heating, is a defining trait of several crystals. Upon irradiation of the crystal, a portion of the absorbed energy is retained within the crystal lattice for extended durations. Heat can induce the crystal to revert to its original state, thus releasing energy in the form of light.

4.16. DOSIMETRY

Many materials, when appropriately doped, can serve as thermoluminescent dosimeters (TLDs). The main TLD is LiF:Mg,Ti, which consists of lithium fluoride doped with magnesium and titanium. Naturally occurring lithium consists of two stable isotopes, ^{6}Li and ^{7}Li, with the latter exhibiting an abundance of approximately 92.5% (Tomascak et al., 2016). A benefit of LiF is that its photoelectric atomic number marginally exceeds that of water. Lithium (Z = 3) possesses a single electron in its valence shell, but fluorine (Z = 9) requires one additional electron to complete its shell. LiF is a cubic crystal composed of ionic bonds between Li^{+} and F^{-} ions. The Mg and Ti dopants are integrated into the crystal structure in complex behaviors. Mg is integrated as Mg^{2+}, leading to a lithium vacancy. The precise mechanism of thermoluminescence is under investigation, but it relies on a dopant, likely Mg++, functioning as an electron trap that captures electrons liberated by ionizing radiation. These traps exhibit high stability with prolonged de-excitation periods. The quantity of trapped electrons is directly proportional to the extent of ionizing radiation received by the crystal. Electrons can be liberated from traps by elevating the temperature of LiF. Some liberated electrons are seized by luminous centers, possibly associated with titanium. Upon capture, these electrons emit their surplus or excess energy in the form of light. The emitted light enters a photomultiplier tube.

CHAPTER 5

PLASMAS AND RADIATION PHYSICS

Contents

5.1. INTRODUCTION

A plasma (Figure 5.1) is produced by supplying energy to a gas, resulting in the release of electrons from atoms and the formation of free electrons and ions (Yasoob et al., 2022). Plasmas respond to applied electric and magnetic fields and alter electromagnetic waves in ways different from those materials.

Figure 5.1. *Plasma.*

Source: https://upload.wikimedia.org/wikipedia/commons/4/40/Plasma_jacobs_ladder.jpg, Creative Commons License.

The distinct behavior of plasmas has led to its classification as a fourth state of matter, alongside solids, liquids, and gases. Over 99% of the observable universe consists of plasma. The Sun is a plasma and constitutes 99.85% of the solar system mass; the proportion of plasma in the solar system increases less when interplanetary plasma is considered. The current understanding of the universe has been facilitated by the observation of electromagnetic radiation released by plasma material.

The fusion of hydrogen isotopes in plasmas, either magnetically restricted or confined or inertia-constrained prior to the expansion of a dense plasma, is expected to offer a novel energy source to replace fossil fuel combustion, while the technical challenges remain unresolved. The fuel for a fusion reactor, deuterium isotope of hydrogen, is abundant in seawater at a concentration of 33 mg/L. Extensive experiments are currently being conducted to create fusion reactors due to the significant potential impact of establishing a fusion power plant. Plasmas are utilized in various technical applications, such as semiconductor etching and thin-film coating. Plasma is generated during the welding of solid materials and is being investigated for biological and medical uses, including bacterial sterilization. The emission of light from plasmas has many applications, including fluorescent tubes and the utilization of ultraviolet light produced by laser plasmas for semiconductor lithography (Figure 5.2).

Figure 5.2. *Semiconductor lithography.*

Source: https://upload.wikimedia.org/wikipedia/commons/e/e3/EB_litograph.jpg, Creative Commons License.

A variety of lasers employing plasmas have been created, including argon ion lasers and a broad spectrum of plasma lasers intended for operation at short wavelengths (5.9 nm). A plasma is characterized as an accumulation of ions and free electrons, wherein the charged particles generate reactions to electric and magnetic fields, while the net charge density averages to zero across extended spatial scales. Similar definitions have frequently been employed to characterize matter in the plasma state. The plasma frequency is vital for understanding the transmission of electromagnetic radiation in a plasma. This results in plasmas emitting electromagnetic radiation that varies with temperature, typically within the infrared to X-ray spectral range, although the emission spectrum may also include longer wavelength microwaves and radio waves, as well as high-energy gamma rays. In plasmas, electrons frequently occupy the excited bound quantum states of ions as well as the free unbound quantum states. This excitation and ionization result in radiation emission. Relatively low-density high-temperature plasmas, atoms and ions can be considered to possess an atomic structure akin to that of isolated atoms or ions, although with quantum-state populations significantly deviating from equilibrium. Free electrons, photons, ions, and atoms undergo collisions with ions, resulting in excitation and ionization. Astrophysical and space plasmas exhibit energy-density ranges from exceedingly low (interstellar space) to exceedingly high (dwarf stars) and are linked to extended timescales, frequently characterized by equilibrium populations and radiation fields.

5.2. PLASMA

The atomic and radiation physics of plasmas includes a spectrum of physics concepts, including electricity and magnetism, quantum mechanics, particle collision theory, relativity, atomic structure, and statistical physics. The investigation of light emission and collision phenomena related to plasmas has yielded substantial support for quantum mechanics. The analysis of light emission and absorption is an efficient and non-invasive technique for assessing plasma properties, including density and temperature. The analysis of spectral emission and absorption is the diagnostic method utilized for astrophysical plasmas and is crucial for assessing conditions in magnetic and inertial fusion plasmas. Light probing techniques, including scattering, absorption, and interferometry, can be employed to ascertain plasma conditions in laboratory settings. An understanding of radiation interaction in plasmas facilitates the interpretation of such probing.

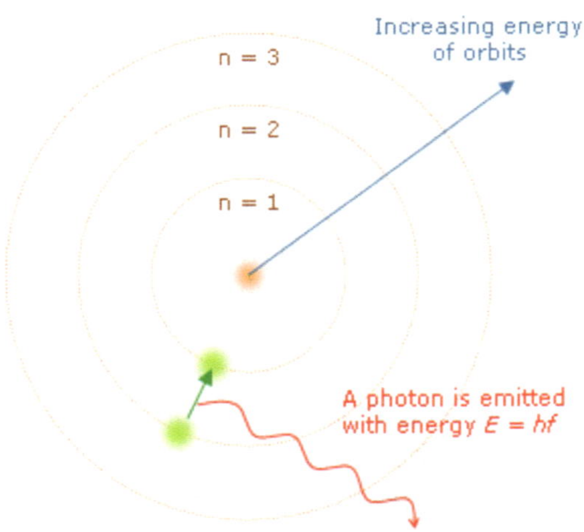

Figure 5.3. *Bohr model.*

Source: https://upload.wikimedia.org/wikipedia/commons/d/d3/Bohr_atom_model_English.png, Creative Commons License.

At elevated densities, electron – ion collisional processes prevail over radiative processes in the excitation and de-excitation between quantum states. Electron-ion collisions result in the transfer of an electron between bound states or, by collisional ionization, facilitate the transition of an electron from a bound state to a free electron quantum state (Müller, 1991). The reverse of this process, in which free electrons collide with an ion resulting in one electron transitioning to a bound state while another electron absorbs the energy from the recombining electron, is termed three-body recombination. The critical matter is that each collisional excitation or ionization event, for a system in thermal equilibrium, occurs at the same rate as the inverse collisional de-excitation or three-body recombination event. The quantum state populations will achieve equilibrium. At low densities, radiative decay rates become prominent relative to electron collisional decay rates; however, the spontaneous decay process is frequently unbalanced by radiative processes like photo-absorption and stimulated emission. This may result in the populations of low-density plasmas deviating from thermal equilibrium populations. The basic atomic system is a hydrogen atom, consisting of an electron and a proton. A slightly complex scenario involving a hydrogen-like ion, consisting of a nucleus with Z protons and one electron, can be addressed using nearly identical equations. Ions possessing an equivalent amount of electrons constitute an isoelectronic series with uniform electron configurations. The basic isoelectronic series is the hydrogen-like series. We can examine a preliminary model of the hydrogen atom that integrates early concepts of quantum mechanics, including the wave-like nature of electrons with an associated wavelength, and accurately predicts the electron energies within atoms.

5.3. BOHR MODEL

The Bohr model (Figure 5.3) exhibits accuracy in many predictions and is consequently often taught to atomic physics students before they delve into more thorough quantum-mechanical representations of atoms. The Bohr model of hydrogen and hydrogen-like ions postulates that an electron orbits a point nucleus with charge Ze in a circular planetary orbit of radius r. The de Broglie wavelength corresponds to the wavelength linked to the wave function of a free electron with a specified velocity. The Sommerfeld extension of the Bohr model partially addressed the angular momentum problem and facilitated an explanation of energy splitting by theorizing that electron orbits are elliptical. The Bohr/Sommerfeld model postulated that the electron traversed elliptical orbits around the nucleus, analogous to a comet's orbit around the Sun (Hentschel, 2018). Relativistic effects became significant as the electron velocity approached the speed of light while traversing near the nucleus. The ellipticity indicated by an alternative quantum number therefore resulted in minor relativistic adjustments to the electron energy. The assumption that elliptical orbits around the nucleus generate energy differences due to varying magnetic fields accounts for the observed energy difference between, for instance, the photon energies of doublet spectral lines. The elliptical orbit model became obsolete with the advent of quantum physics and the realization that the notion of a free electron particle orbiting in circular or elliptical paths is merely an approximation. However, we must recollect the correspondence principle before proceeding to examine quantum-mechanical solutions for atoms. As the density of quantum states increases, quantum mechanical solutions converge towards classical theory. For high primary quantum numbers (n), the energies of the circling electrons converge, varying inversely with the square of n, as per the Bohr model. The high n electron states increasingly resemble the Bohr model, where an electron orbits the nucleus in a classical trajectory. For an atom with atomic number Z, a high n electron perceives a charge around the nucleus consisting of Z protons and (Z − 1) electrons, resulting in the core atomic potential resembling that of a hydrogen nucleus. Atoms possessing an electron in higher n orbits are designated as Rydberg atoms, as they exhibit classical orbital behavior as delineated by Bohr, with their energies determined by the Rydberg energy divided by n^2. Electron transitions with higher n values result in radio-wave absorption or emission, making Rydberg atoms significant in radio astronomy. The Bohr/Sommerfeld model, which depicts electrons following elliptical orbits around the nucleus, has similarity to the quantum mechanical distribution of wavefunctions surrounding the nucleus. This closeness is anticipated, especially for higher n states, according to the correspondence principle. We might anticipate a similarity, as the Bohr–Sommerfeld model precisely predicts atomic energies, including fine-structure energy levels that result in phenomena such as spectral-line doublets, whereas the basic Bohr model predicts a singular spectral line. Generating a plasma with unbound electrons needs high temperatures, making light emission, propagation, and absorption significant factors. The transmission of light, in contrast to many other waves, does not require a medium for oscillation. Light propagates through plasma, causing the free electrons to oscillate; however, these electrons typically delay wave oscillation rather than facilitate it. Light cannot propagate in plasma if the radiation frequency is lower than the plasma natural oscillation frequency, as elaborated. At low frequencies, electron oscillations in relation to the ions attenuate the electromagnetic oscillation of light.

5.4. LIGHT

Light of all frequencies crosses a vacuum without absorption; however, the intensity from any source diminishes in proportion to the inverse square of the distance from the source. The acceleration of any charge generates transverse electric field oscillations, hence establishing a mechanism for the generation of electromagnetic waves. The electromagnetic spectrum relevant to plasma physics extends from radio waves to X-rays and gamma rays. The transmission of all components of the electromagnetic spectrum entails the same physical principles, with deviations arising during interactions with matter. The highest frequency, highest-energy gamma rays, with photon energies above 100 keV, are generally not produced by thermal processes; however, they can be significant when rapid superthermal particles are generated in plasmas. Radio-wave propagation is significant in low-density plasmas, such as the ionosphere. Certain phenomena of electromagnetic wave propagation in plasma manifest at low irradiance but become predominant in energy transfer and other processes at the high irradiances associated with laser light. Concentrated high-power laser light directed at a solid surface or into a gas generally generates free electrons within the initial oscillations of the electric field, owing to the substantial electric field linked to even moderate laser intensities. In laser-solid target interactions (high-power), the majority of a laser pulse lasting more than a few femtoseconds engages with plasma expanding nearly perpendicularly to the target surface. High irradiance is generally characterized by irradiances above 10^9 Wcm^{-2}, at which point substantial ablation and ionization of solid materials or the ionization of gases occurs (Naik et al., 2017). With some targets, plasma material expands uniformly to the target surface, and the light interacts with a planar expanding plasma. The light can infiltrate the plasma to the critical density. A similar shape or geometry for radio waves interacting with the almost planar plasma of the ionosphere arises at a significantly lower critical density. The ionosphere is a higher layer of Earth atmosphere that becomes ionized by solar radiation during day hours. It is situated 75 to 1,000 km above the Earth surface, with a maximum electron density at roughly 160 km. We can examine the propagation path of radiation in a planar plasma, which possessing high temperature is crucial for understanding the interaction of radio waves with the ionosphere. The difference in mass between electrons and ions results in an electromagnetic wave interacting primarily with the electrons in a plasma, leading to a negligible direct acceleration of the ions. We can initially analyze the dynamics of an individual free electron within an electromagnetic wave. We can assume that the electromagnetic wave is linearly polarized and propagates in the z-direction, with the electric field oriented in the x-direction. The magnetic field of the electromagnetic wave is oriented in the y-direction. At higher laser irradiances, the photon density is adequate for many photons to collectively ionize one electron from a bound atomic state or generate an electric field strong enough to alter the atomic potential, facilitating ionization. These processes generate the free electrons when a high-power laser engages with a substance, such as a solid surface or a neutral gas. The energy from the absorbed photons must exceed the combined total of the ionization potential and the ponderomotive potential (Dörner et al., 2002). At irradiances where the ponderomotive potential begins to inhibit direct multi-photon ionization, the perturbation of the atomic potential by the electric field of the electromagnetic wave becomes predominant, allowing electrons to either tunnel through a lowered or reduced barrier or possess adequate energy in their bound state to overcome the diminished potential barrier. Following the initial ionization via multi-photon processes that generate the first few free electrons, collisions between free electrons and ions predominantly facilitate ionization throughout the majority of the laser pulse duration impacting a target. Analysis of electromagnetic propagation in a plasma presumed the plasma to be homogeneous. Plasma generated by laser irradiation of a solid generally expands perpendicularly to the initial target surface, since the aspect ratio of the initial plasma material spans a laser range of 5–100 μm, whereas the depth of the ablated material is usually around 1 μm. The initial plasma is thin, resulting in expansion mostly along the direction of the least thickness, which aligns with the target normal direction. An assumption of a flat target surface spanning the size of the focal spot is generally a valid approximation, even for spherically shaped targets. In laser-generated plasmas, the Debye length is generally significantly lower than the plasma dimensions, resulting in ions and electrons expanding concurrently, with fast ions traveling further. A plasma expansion can be rapidly produced when faster ions travel further from the target. The profile displays a linear velocity gradient relative to the distance from the target, but the density profile decreases approximately exponentially with increasing distance from the target surface. At a certain distance from the target surface, the electron density falls below the critical threshold, marking the closest penetration distance of any incident laser light to the target surface. The interaction between an electromagnetic wave and an increasing electron density was initially examined regarding the propagation of radio waves into the ionosphere prior to the start of the laser. In resonance absorption, electrons are propelled by an intensified electric field, enabling them to attain high energies without colliding with ions or other electrons, as collision cross-sections diminish significantly for high-velocity particles with increasing particle velocity. Superthermal electrons are frequently generated, possessing a high temperature than the thermal temperature of the majority of the plasma. Superthermal electrons can transfer energy at distances from the laser focus or result in energy loss from the plasma as they disperse. In inertial fusion, superthermal electrons provide significant challenges as they can cross considerable distances and pre-heat the deuterium–tritium fuel, complicating the compression process.

5.5. EMR

The emission and absorption of electromagnetic radiation occur when charges are accelerated. A charged object of any mass emits radiation when accelerated; yet, electrons are the emitters and absorbers due to their low mass, which facilitates superior acceleration in plasmas and other mediums, such as conducting wires. All electromagnetic radiation emission and absorption result from the acceleration of charged particles. For particles accelerating at rates nearing the speed of light, relativistic effects become significant. We can gain a qualitative understanding of the implications of relativity by examining how the transformation from the particle frame of reference to the rest frame influences the behavior. When the charge moves at a constant high velocity nearing the speed of light prior to the acceleration, the electric field disturbance relative to the charge is asymmetric for positions aligned with the initial velocity and those opposite to it in the rest frame. In the rest frame of reference, distances for electric field disruption are compressed along the direction of motion and elongated in the direction opposite to the velocity.

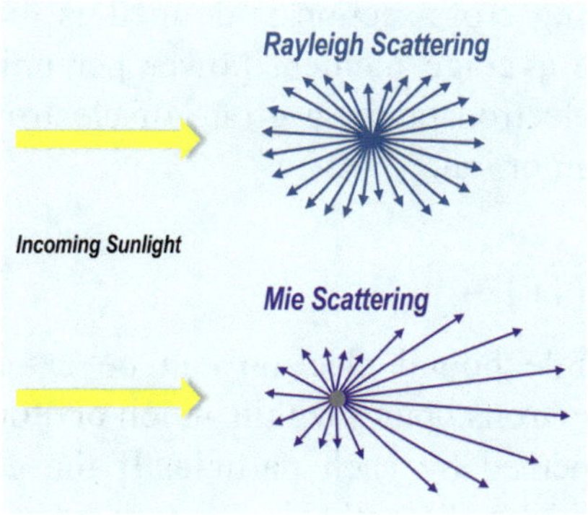

Figure 5.4. *Mie scattering.*

Source: https://pace.oceansciences.org/images/scattering_What_is_PACE_v3.jpg, Creative Commons License.

The overall result is that the power emitted in the rest frame is amplified in the direction of the velocity and diminished in the opposite direction. The acceleration of electrons influences the propagation of electromagnetic waves and results in radiation emission; so, the scattering of light by electrons in a medium can be considered a cause of the medium optical properties. The resonances in the reactions of free and bound electrons to oscillations from electromagnetic waves significantly influence light propagation. We can calculate the refractive index in plasmas due to free electrons and the resonance at the plasma frequency. Other resonances related to bound electrons also generate refractive index effects. The refractive index of the medium through which light travels dictate the scattering processes that eventually influence the reflection and refraction of light at the interface between materials with varying refractive indices (Meyzonnette et al., 2019). Macroscopic particles like dust in plasmas or water droplets in atmospheric clouds reflect light from surfaces, a phenomenon referred to as Mie scattering (Figure 5.4). Gradients in the refractive index result in the bending of light propagation direction. The proportion of electromagnetic radiation scattered by free electrons generally represents a minor loss mechanism for radiation frequencies significantly exceeding the plasma frequency, however it serves as a valuable diagnostic tool for plasma conditions. In diagnostic measurements of plasmas at optical frequencies, a laser radiation source is typically utilized to spatially locate light, and the higher laser power per unit area guarantees that the scattered light surpasses the emission linked to the plasma thermal energy. High-power radar systems utilize radio waves and ionospheric scattering. In dense plasmas related to inertial fusion, incoherent X-ray sources or free-electron laser sources are employed. The scattering cross-section is defined as the ratio of the average scattered power to the average incident power per unit area, and it can be assessed for a single electron or for a group of electrons, such as all bound electrons within an atom or ion.

5.6. SCATTERING

The method for a single bound electron can be extended to consider the scattering of several electrons bound within an ion or atom. We must aggregate the electric fields dispersed by each particle. If the electric fields of each scattering electron combine coherently, the scattering rate escalates well beyond that of an equivalent number of individual electron scatterers, increasing at a rate proportional to the square of the number of electrons scattering coherently. If Thomson scattering is incoherent, with each electron sufficiently separated such that the scattered electric fields do not overlap, the frequency of the scattered photons can be Doppler shifted in proportion to the individual electron velocities. Thomson scattering of light from a moving electron results in a frequency shift attributable to the Doppler effect (Figure 5.5).

We can derive the Planck radiation law for an equilibrium radiation field by examining the photon density within a black-body cavity. The concept of a radiation field in thermal equilibrium is relevant to several astrophysical and certain laboratory plasmas, facilitating the deduction of correlations between inverse processes interacting with a radiation field. When radiation and electrons in a medium undergo many interactions via emission and absorption, the radiation field attains thermal equilibrium and can be considered to possess a temperature equivalent to that of the electrons. An equilibrium radiation field is a set of photons that adhere to the principles of statistical mechanics applicable to a collection of bosons. We must compute the density of modes for light in a steady-state system.

Figure 5.5. *Doppler effect.*

Source: https://upload.wikimedia.org/wikipedia/commons/d/d4/Velocity0_70c.jpg, Creative Commons License.

The Planck radiation energy density is termed black-body radiation due to initial investigations into the absorption and emission of infrared radiation by Gustav Kirchhoff and others. Kirchhoff discovered that a black object serves as an absorber and emitter of infrared light, coining the phrase black body (Khene, 2021). A black body at a temperature slightly above room temperature can emit a Planckian spectrum of radiation, with a peak emission in the infrared range and minimal radiation in the visible spectrum. The intensity and flux of radiation, which quantify power per unit area, are significant near the boundaries of the black-body cavity or when an item is situated within the cavity. For angular variations of intensity and flux, we utilize steradian units, which are defined as the solid angle subtended by an area of a sphere with a unit radius at the sphere center. If the area includes the entire surface of the sphere, corresponding to all angles, the solid angle is 4π steradians. The interaction of radiation with plasma atoms and ions includes a lower and upper quantum state. Our understanding of the Bohr model for hydrogen and hydrogen-like ions is adequate to consider an energy level framework for electrons.

5.7. EINSTEIN EXPERIMENTS

Einstein initially proposed the concept of stimulated emission, wherein a photon interacting with an excited atom induces the atom de-excitation and the emission of a photon identical to the stimulating photon. For many years following Einstein's proposal of stimulated emission, it is mostly utilized in models of radiative processes. This transformation occurred with the introduction of the laser in 1960 and the subsequent development of applications utilizing the collimated, and coherent radiation produced by lasers. The emission of radiation is a crucial tool for detecting plasma conditions and can significantly contribute to energy transfer. In plasmas, even at the highest concentrations, electrons can be considered either in bound quantum states, where their energies are mostly influenced by the core potential of ionic nuclei, or unbound, occupying a continuum of free-electron states. It has been shown that free electrons do not occupy an actual energy continuum; yet, a free electron implies a high density of quantum states per unit energy, allowing us to frequently regard free-electron energies as continuous. Radiation emitted by free electrons predominates in low atomic number plasmas. Radiation transition probabilities increase significantly with atomic number, scaling as Z^4 for hydrogen-like ions, whereas radiation for free-electron transitions scales at a lower rate proportional to Z^2. In total thermal equilibrium with the radiation field corresponding to particle temperatures, the emission of radiation is described by the Planck black-body equations. Complete thermal equilibrium is rare in laboratory plasmas and the less dense astrophysical plasmas due to minimal radiation absorption inside the plasma dimensions. A tokamak is a device that employs a magnetic field to confine plasma in a toroidal configuration (Ariola and Pironti, 2016). Coils conducting electricity encircling the torus, in conjunction with an induced electron current traversing the torus, generate helical magnetic field lines oriented around the torus. Charged particles revolve around the magnetic field lines, so being partially constrained inside the magnetic field topology. The fusion of deuterium and tritium isotopes of hydrogen in confined plasma presents significant potential for energy production. An electrical coil situated in the core aperture of the torus functions as the primary coil in a transformer setup, while the toroidally structured plasma serves as the secondary coil. The current-carrying coils encircling the torus generate the primary magnetic field and are positioned at a radius from the center of the central hole to create a magnetic field perpendicular to the radius. In bremsstrahlung, an electron emits electromagnetic radiation while being accelerated in the vicinity of an ion. Inverse bremsstrahlung occurs when an electron within an ion field absorbs energy from an electromagnetic wave. In an equilibrium radiation field, the rates of bremsstrahlung and inverse bremsstrahlung are in balance. We can derive an expression for the absorption coefficient of inverse bremsstrahlung based on the power emitted by bremsstrahlung. An electron can be radiatively trapped by an ion, resulting in the electron becoming bound in a quantum state of the ion. The total kinetic energy of the electron, in addition to the energy differential between the final quantum state and the ionization energy of the ion, is emitted as photon energy. The emission is referred to as recombination radiation or free-bound radiation, as it occurs when electrons transition from free, unbound states to bound, discrete quantum states.

5.8. OPACITY

Radiation opacity in plasma is important at high densities or within extensive plasmas. However, low-density plasma may exhibit some radiation frequencies, particularly within the microwave or radio frequency spectrum ranges, where absorption of radiation is considerable. Plasma opacity can restrict the diagnosis of conditions within a plasma interior and can impede energy outflow, resulting in non-local radiation heating. Radiation transport calculations employ a dimensionless distance metric referred to as optical depth. The optical depth of a material thickness is the natural logarithm of the ratio of incident to transmitted radiation power through that medium. In the absence of any radiation source, the radiation power decreases exponentially with optical depth. When the optical depth nears 0, a medium is considered optically thin for the given radiation, rendering the opacity negligible. High levels of optical depth for a plasma or other medium indicate that opacity and radiation transmission are substantial, a situation referred to as optically thick. Following investigation of opacity, we can analyze plasmas that are in or near thermal equilibrium, allowing the radiation field to be characterized by the Planck black-body distribution. The radiation intensity in a plasma transfers energy to and from bound and free electrons through emission and absorption processes (Boulos et al., 2015). Electron transitions between discrete bound quantum states linked to the energy levels of an ion or atom result in spectral-line emission, whereas transitions from free electrons to other free quantum states or to bound quantum states yield continuous emission. The emission processes can be characterized by an emission coefficient, which denotes the radiated power per unit volume across a frequency range from ω to $\omega + d\omega$. Radiation can be absorbed by electrons associated with ions or atoms, or by free electrons, leading to the conversion of photon energy into kinetic energy or the excitation of an electron to a higher quantum state. The radiation absorption rate can be denoted by an absorption coefficient K, which has dimensions of inverse length. Many radiative transfer problems can be modeled using a geometry. The Bohr model provides an accurate representation of the energy levels of hydrogen and hydrogen-like ions. This model is also a valid representation of the energy of excited states in higher-atomic-number atoms and ions. Excited electrons in multi-electron atoms and ions occupy orbits at a distance from the nucleus and other electrons, resulting in their kinetic and potential energy being influenced by the localized net charge near the nucleus, similar to hydrogen-like ions. The Bohr model, however, does not accurately predict the angular momentum of spinning electrons nor the fine structure of energy levels. The fine structure induces the splitting of energy levels, resulting in the formation of two or more closely spaced spectral lines instead of a singular line. The Bohr model fails to provide an adequate approach for assessing the rates of radiative absorption and emission, a task achievable through time-dependent quantum mechanics. The impact of electron spin energy in the magnetic field generated by electron orbital angular momentum on the energy levels of hydrogen and hydrogen-like ions has been examined. The interplay between spin and orbital angular momentum results in the fine-structure splitting of energy levels. Atomic nuclei possess spin and generate an extra energy resulting from the interaction energy between the nuclear magnetic moment and the magnetic field produced by the electron angular momentum. The energies resulting from nuclear magnetic moment and electron orbital effects are around 1,000 times less than those arising from electron magnetic moment and electron orbital effects; so, the minor energy level fluctuations caused by nuclear spin are referred to as hyperfine structure. Multiple electron atoms possess electrical configurations similar to those of the hydrogen atom; however, the shielding effect of the nuclear potential by other electrons results in electrons with lower $l=0$ quantum numbers being more closely bound, hence exhibiting lower energy levels. The Pauli exclusion principle dictates that a single electron may occupy each electronic quantum state exclusively (Pauli, 1994). The ground state of elements is established by populating each electronic quantum state according to the quantity of electrons corresponding to the atomic number. Multiple electron atoms with complete (full) n and l quantum numbers, referred to as sub-shells, exhibit spherical electron probability distributions, possess zero angular momentum, and have no magnetic moment. The Coulomb repulsion among electrons results in a structure of more intricate energy levels. If the electron probability distributions are farther distant or apart, the Coulomb repulsion diminishes, resulting in stronger electron binding and a reduced overall atomic energy. A greater total spin S and a larger total angular momentum L result in individual electron probability distributions that are more dispersed, hence leading to lower energy levels. The spin-orbit coupling energy between the magnetic field linked to the total angular momentum of the atom and the total electron spin results in additional quantum state energy partitioning into multiplets. A molecule is defined as a cluster of two or more atoms bonded through the sharing of one or more electrons. A covalent bond is formed when there is almost equal sharing of electrons between atoms. The shared electrons in a covalent bond attract the positively charged atomic nuclei enough to exceed the mutual electrostatic repulsion between the nuclei. When the atoms of a molecule consist of different elements, the electrons may not be distributed uniformly around the nuclei. The attraction between atoms can vary based on the extent of electron sharing, resulting in a net negative charge near one atom and a net positive charge near another, thereby forming an ionic bond.

5.9. ELECTRON BEHAVIOR

Upon populating electrons into sub-shells of increasing energy, the final unoccupied sub-shell predominantly governs the chemical behavior of the element, as the wave function of the electron in this sub-shell extends further from the nucleus, thereby increasing the likelihood of interaction with adjacent atoms. Neutral and ionized molecules exist in plasmas when particle temperatures are low enough that vibrations and collisions lack the energy required for molecular dissociation. The plasma ions interact with surfaces or other plasma components to generate molecules. Temporary molecules are occasionally generated in plasmas. Charge exchange occurs when a neutral atom, such as hydrogen, combines with a highly ionized atom to create a temporary molecule. Research indicates that to attain optimal accuracy, the Rosseland mean opacity must incorporate molecular transitions at temperatures under 5,000 K. We can examine a basic molecular ion consisting of two protons and one electron: the hydrogen ion molecule. The physics of the H_2^+ molecule exemplifies the behavior of more intricate molecules and is pertinent to plasma physics. The ionization of neutral molecular hydrogen gas produces the hydrogen ion molecule. Molecular H_2^+ is generated in sources utilized for producing neutral beams of atomic hydrogen, which are employed to heat tokamak plasmas (Fantz et al., 2021). The Saturn magnetosphere is known to contain H_2^+ ions. Ground and excited electron states have identical forms, with the minimum energy increasingly shifting towards greater internuclear distances for higher energy excited states. If the atoms constituting a molecule are dissimilar, the least energy configuration is not invariably achieved when the wave function is uniformly divided across the two nuclei. An ionic bond occurs when the electron wave function exhibits lower energy when localized near one atom, resulting in a net negative charge, while simultaneously generating a net positive charge in the zone of the other atom. At non-zero temperatures, the internuclear lengths between the atoms of a molecule are variable and exhibit temporal oscillations. Molecules are permitted to rotate freely in space. The negligible masses involved result in significant quantum mechanical quantization of motions, yielding discrete vibrational and rotational energy levels. We can choose the Born-Oppenheimer hypothesis, which postulates that nuclear mobility is isolated from electron dynamics due to the significantly higher mass of the nuclei, allowing electrons to rapidly adapt to any changes in nuclear positions. The Born-Oppenheimer approximation permits the electron wavefunctions and energy eigenvalues to remain independent of nuclear velocities. The overall energy of a molecule is the sum of the electron energy and the reduced quantized vibrational and rotational energies of the nuclei. Radiation can alter the energy state of matter by three optical processes: spontaneous emission, absorption, and stimulated emission.

In spontaneous emission, an excited atomic or molecular state transitions to a lower energy level, emitting a photon whose energy corresponds to the difference between the quantum state energies. Photon absorption is the inverse process, wherein a photon induced excitation from a lower to a higher quantum state. Stimulated emission occurs when an excited quantum state is prompted to transition to a lower quantum state by an incident photon. An existing photon persists while an identical photon is generated, thereby conserving energy.

5.10. QUANTUM MECHANICS

The frequencies of radiative transitions can be determined by quantum mechanics (Freed, 2002). The time-dependent Schrödinger equation is employed with the wave function consisting of time-varying combinations of the two wavefunctions corresponding to the lower and upper quantum states. In semi-classical approaches, an atom is analyzed using quantum mechanics while the radiation field is addressed differently. A disturbance in energy arises from the contact between the electric field of radiation and the atom electric dipole, however other interactions may become relevant if the electric dipole interaction energy is minimal. The theory of relativity establishes the correlation between the frequency of radiation as perceived by an observer and that emitted by the source. At elevated densities, the energy levels of an ion within a plasma or the radiation process itself are influenced by the surrounding environment. The phenomenon affecting spectral line profiles is termed pressure broadening; however, the terms Stark broadening or collisional broadening are employed when the primary contribution to the broadening process stems from quasi static energy level shifts induced by the Stark effect from micro-electric fields generated by nearby ions in the plasma or collisions between ions or electrons and emitting ions. These processes manifest concurrently; yet, similar to other line broadening phenomena, a single mechanism may predominate for a certain spectral line. Pressure broadening of spectral lines occurs due to the influence of the plasma environment on the emitting ions. The immediate environment of an ion in a plasma fluctuates, resulting in a broadening of the ensemble average line-profile emission. The emission process or the energy of quantum states are altered due to the perturbing action of adjacent ions or electrons. A comprehensive strategy to line broadening can be developed that, in theory, includes all pressure-broadening effects simultaneously. When the emission process occurs from an ion that remains effectively stationary during the emission period, the predominant perturbing influence is often the electric field generated by adjacent ions. The energy levels are altered based on the micro-electric field specific to each ion, resulting in ensemble line broadening. The impact is particularly significant when energy levels degenerate without an electric field, but become split due to the Stark effect in the presence of a field. Collisional broadening occurs when the exponential decline of the electric field over time, due to a transition between two confined quantum states, is disrupted by a collision with another ion or electron interacting with the emitting ion. In an electron-ion collision, the primary consequence is to induce a collisional transition from the ionic upper quantum state, hence diminishing the effective span of that state. In ion-ion collisions, the colliding ion disrupts the energy levels of the emitting ion, hence interrupting the electromagnetic emission wavetrain. The cumulative impact of both electron-ion and ion-ion collisions diminishes the duration of coherent emission from an atom involved in a collision.

5.11. COLLISIONS

Ion-ion collisions may induce a spectral narrowing of the line width for Doppler-broadened lines (Stambulchik and Maron, 2006). The ion-ion collisions occur at a significantly accelerated rate compared to the radiative emission timescale. The effective average velocity reduction resulting from the swift directional changes of the emitting ions due to collisions during the average emission time indicates that the average Doppler frequency shift is diminished below the value calculated from the instantaneous ion velocity derived from the ion temperature. The phenomenon is termed Dicke narrowing, (Robert Dicke), who first demonstrated its influence on line profiles. The measurements of spectral line widths may be influenced by additional factors. The spectrometer employed for linewidth measurement will possess a wavelength resolution and an instrument profile. A Doppler shift of the line profile occurs if the plasma being observed has a relative velocity to the observer. Line profiles may be simultaneously expanded by multiple methods. The spectrometer (Figure 5.6) measuring the profile may possess a resolution similar to the line width, leading to a measured line profile shaped by both the instrument resolution and the emitted profile.

Collisions among electrons, ions, and atoms in a plasma result in the exchange of energy and momentum, leading to the formation of a thermal distribution characterized by temperature and density. We can examine the concept of an electron colliding with an ion, wherein the electron acceleration results in bremsstrahlung emission. A minimal energy loss to radiation occurs with each impact. In collisions involving particles of similar mass, such as electron–electron and ion–ion interactions, momentum and energy can be easily transferred between the charged particles. Collisions between electrons and ions also transmit energy, although at a rate that is diminished by the relative mass of the ions to the electrons.

Figure 5.6. *Spectrometer.*

Source: https://upload.wikimedia.org/wikipedia/commons/a/a0/Silver_Target_in_XPS_Spectrometer.jpg, Creative Commons License.

A collision is termed elastic when the kinetic energy of the colliding particles is preserved, excepting minor bremsstrahlung emissions. Inelastic collisions arise in alterations to the bound quantum state of an ion or atom due to the collision. Energy is transmitted between the kinetic energy of an electron and the potential energy of a bound electron or vice versa. A critical parameter in all collisions is the cross-section value for the collision. The cross-section possesses area dimensions and can be conceptualized as the region surrounding the ion or atom that defines the requisite proximity for another colliding particle to induce a specific alteration in the ion quantum state, in the case of inelastic collisions or a deviation of the colliding particles, in the case of elastic collisions. To assess collisional cross-sections between electrons and ions, one must account for the incoming electron wave function and its interaction with the central potential arising from the nucleus charge and the bound electrons. A method for assessing the inelastic collisional excitation and ionization cross-sections, in which the quantum state of the ion varies, could utilize the time-dependent Schrödinger equation. The interaction between an electron and an ion during a collision is influenced by the spatial change in the distance between the electron and the ion. The interaction is effectively analyzed by examining the spatial variation of the wave function as it interacts with the ion, utilizing the time-independent Schrödinger equation. The free electron can be viewed as an extended wave-train that creates a stable interference pattern as a result of the ion presence. In the analysis of elastic collisions inside plasmas, we can focus on effective point charges, such as electrons or the central potential of ions, leading to the incident particle experiencing a Coulomb potential from the colliding particles. At moderate to high electron concentrations, three-body recombination has a large cross-section, as both momentum and energy can be conserved during the collision. At diminished electron densities, the likelihood of two electrons participating in a three-body recombination collision with an ion decreases, and there is an absence of a third or non-recombining electron to facilitate momentum exchange during the impact. In the low electron densities present in the ionosphere and laboratory low temperature plasmas, electron-ion recombination is frequently governed by the dissociative recombination of molecular ions (Hillier, 2023). During an electron-molecular ion collision, an incoming electron experiences repulsion from the bonded electrons within the ion and attraction from the nuclear charge. The incoming electron can occupy an electronic state of the ion and stimulate a bonded electron, preserving energy. The doubly excited state of the neutral molecule is typically non-bonding, resulting in the dissociation of the molecule into individual atoms. The incoming electron may occupy a highly excited electronic state of the molecular ion, resulting in a highly excited state of the neutral molecule. The highly excited neutral molecular state possesses a significant likelihood of dissociation due to elevated vibrational energy. An alternate pathway is auto-ionization, resulting in a molecule ion and a free electron. In a plasma at thermodynamic equilibrium, the radiation field adheres to the Planck black-body formula, ionization is governed by the Saha-Boltzmann equation, and the population ratios of bound quantum states are dictated by the Boltzmann ratio. In local thermodynamic equilibrium (LTE), quantum state populations are determined by the Saha-Boltzmann equation and the Boltzmann ratio, while the radiation field is not in equilibrium with the particles. It is important to highlight that local thermodynamic equilibrium (LTE) often arises when collisional events predominantly influence the population dynamics of quantum states, while radiative processes remain negligible. The cross-sections depend on the energy of the incident colliding electron; however, in a plasma, there exists a Maxwellian distribution of the energies of the free electrons.

5.12. MAXWELL DISTRIBUTION

The cross-section values must be averaged over the Maxwellian distribution to get a rate coefficient, which, when multiplied by the density of free electrons and the initial quantum state density, results in the rate of change of the quantum state. Models that compute plasma quantum state densities and the resulting radiation emission and absorption characteristics through the rates of radiative and collisional processes are referred to as collisional radiative models. A spectral line, referred to as a dielectronic satellite line, is generated from a plasma when two electrons occupy excited quantum states. The energy levels of the quantum states are influenced by the presence of the spectator electron, but to a lesser extent than if the spectator electron were in the ground state. The energies of the doubly excited or dielectronic quantum states are slightly reduced from the energies of the ion with the spectator electron entirely removed, corresponding to the subsequent ionization stage. The dielectronic lines serve as satellites to the lines of the subsequent ionization stage, displaying at higher wavelengths. The abundance of doubly excited quantum states allows dielectronic recombination to substantially influence the ionization level in plasmas. Auto-ionization is the reverse process of dielectronic recombination. Auto-ionization, sometimes referred to as Auger decay, typically denotes the non-radiative decay of excited neutral atoms (Mehlhorn, 1983). Complete thermodynamic equilibrium is achieved when the populations of ion quantum states are balanced and the radiation field is in synchronization with them. In the presence of many interactions, the radiation field becomes Planckian, the ionization equilibrium is described by the Saha-Boltzmann equation, and population density ratios are determined by the Boltzmann ratio. In LTE, the distributions of quantum states are described by the Boltzmann ratio, whereas the ionization degree is ascertained using the Saha-Boltzmann equation. The radiation field typically lacks equilibrium, meaning it does not conform to a Planckian black-body distribution. In LTE, the plasma may be optically thin to emitted radiation. In LTE, equilibrium populations are attained as collisional processes prevail over radiative processes. Inverse collisional processes exhibit balance. Therefore, in the absence of any additional processes beyond the balance, the populations attain equilibrium values corresponding to the specific density and temperature of the plasma. Significant energy quantum transitions predominantly take place in lower-energy quantum states, while the higher-energy states, which are more densely packed, have populations influenced by collisional excitation and de-excitation mechanisms. The balanced collisional processes, coupled with minute radiative decay, result in the higher energy states, which are closely spaced, achieving equilibrium, with their populations dictated by the Boltzmann ratio.

5.13. PLASMA DENSITY

Plasma density can be measured by spectral line emission ratios when the influence of electron density on one spectral line differs from its effect on another (Bastiaans and Mangold, 1985). In a plasma exhibiting coronal equilibrium, quantum states that are metastable against radiative decay experience an increase in population, resulting in emission that is governed by the rate of collisional excitation. The collisional excitation rate can be considerable, although the radiative decay rate is minute. As density increases, collisional transitions to other excited quantum states and the ground state begin to deplete the excited metastable state. At a certain density, the population of the metastable state will be described by the Saha-Boltzmann equation, exhibiting a significantly diminished relative population and a reduced spectral line emission compared to that of a non-metastable quantum state. The transition from coronal metastable emission to diminished relative emission in local thermodynamic equilibrium is based on density, such that comparing the intensity of the metastable line with a reference of a highly emitting allowed transition reveals significant variations in electron density. In the emission spectra of helium-like ions with moderate to high atomic numbers, metastable quantum states exist. The radiative decay to the ground state is metastable, as angular momentum conservation necessitates a change in electron spin during the transition, significantly diminishing the radiative decay rate. The requisite density for achieving a high-density plasma likely varies. The equilibrium relationships for ionization and radiation distribution were employed to infer rates of inverse processes by applying detailed balance. At high densities and reduced temperatures, it is essential to amend the equilibrium ionization relation (the Saha-Boltzmann equation) to accommodate the complete occupation of free-electron quantum states, necessitating the use of the Fermi-Dirac distribution in place of the Maxwellian electron energy distribution. At higher densities, photons exhibit an increased propensity to interact with particles. The Planck black-body radiation distribution remains unaltered at higher photon or particle densities, as photons are bosons and an infinite quantity can occupy an excitation state. In conditions of high density and low plasma temperature, the chemical potential and Fermi energy related to the almost complete occupancy of free-electron quantum states are significant. The energy linked to free electrons with almost complete occupancy of free-electron states generates a degeneracy pressure, since electrons are unable to occupy low-energy filled states and necessitate additional energy to occupy vacant higher-energy levels. Pressure shares the same dimensions as energy density per unit volume, and even at higher densities, a plasma frequently approximates an ideal gas, wherein the numerical value of gas pressure equals two-thirds of the energy density per volume.

At exceptionally high densities, the presumption that ions possess quantum states primarily influenced by the core nuclear potential begins to weaken. Atomic and radiation physics frequently govern the energy dynamics and physical state of plasma materials at high densities where LTE or near-LTE conditions prevail (Scott, 2005). High-density plasma is generated when a material is subjected to laser heating. Laser-generated plasma densities vary from exceeding solid density to extremely low densities as the plasma expands into a vacuum, approaching the densities attained by the vacuum pumping system. Lasers functioning in the infrared to ultraviolet spectrum engage with high-density plasma originating from a solid surface. The initial free electrons are generated through multi-photon absorption linked to photons at the start of the laser pulse. When a significant quantity of free electrons is available, laser absorption occurs via inverse bremsstrahlung and additional mechanisms such as resonance absorption. Energy is transmitted from the critical density, where laser light penetrates into the solid material, resulting in the formation of high-density, low-temperature plasma. All laser-generated plasma material originating from a solid transitions through a high-density, low-temperature state throughout the ablation process. The plasmas generated by concentrating X-ray and ultraviolet lasers on targets can produce plasma with high-density thickness and homogeneity compared to that obtained with concentrated infrared to ultraviolet lasers. X-ray and ultraviolet lasers have been constructed utilizing plasma media and free electron lasers, wherein linear particle accelerators generate lasing activity through the oscillations of electrons in alternating magnetic fields, referred to as undulators. This recent technology has enabled the exploration of a novel domain in high-density plasma physics. The frequencies of X-ray laser light must exceed the plasma frequency generated at solid density, resulting in the absence of reflection or absorption. The plasma generated by an X-ray laser initially absorbs the laser radiation through photo-ionization and subsequently through both photo-ionization and inverse bremsstrahlung. A significant portion of the laser energy is attributed to direct photo-ionization, resulting in very low electron temperatures (100 eV). In direct photo-ionization, electron temperatures may be minimal, and since pulse lengths are typically less than 100 fs, ion temperatures are relatively low, around room temperature, during the laser interaction. The compressed plasmas examined for inertial fusion demonstrate an extreme of high-density plasma. Inertial fusion, or inertial confinement fusion (ICF), aims to generate energy by compressing a deuterium and tritium fuel target to high density, facilitating fusion processes. Compression is accomplished either through the generation of X-rays in a cylindrical black-body (hohlraum) cavity via laser heating or through direct laser irradiation of a fuel pellet. The deuterium-tritium fuel is generally solidified within a spherical plastic pellet. X-ray or direct laser ablation of the fuel pellet shell generates fuel compression due to the conservation of momentum from the external momentum of the ablated material. In inertial fusion, the plasma is not contained by magnetic fields or alternative methods, so an optimal situation involves the complete reaction of deuterium and tritium fuel prior to the cessation of fusion events due to plasma expansion. At the core of a compressed spherical pellet, a minor hot spot of material experiences a 10^3-fold increase in density, initiating the initial fusion reactions that yield helium nuclei (α particles) with a kinetic energy of 3.5 MeV and neutrons with an energy of 14.1 MeV. The α particles subsequently heat an inner layer of the fuel, resulting in additional fusion and the formation of a small expanding spherical shell of material undergoing fusion, referred to as a burn-wave. The burn wave must travel outward at an adequately rapid rate to ensure that all the fuel achieves fusion prior to substantial expansion. In inertial fusion, the deuterium-tritium fuel is preferably maintained at a relatively low temperature during compression to optimize the process. Free-electron degeneracy effects on plasma pressure are significant for cold, dense plasmas. Inertial fusion research has gained advantages from the intersection of certain physics principles with processes relevant to nuclear weapons; yet, it has also faced limitations due to uncertainties around nuclear proliferation.

CHAPTER 6

NUCLEAR MEDICINE AND RADIATION PHYSICS

Contents

6.1. INTRODUCTION

Single-photon emission computed tomography (Figure 6.1), or SPECT, cameras capture several planar images of the radioactivity within an organ (Alqahtani, 2023). The data are subsequently subjected to mathematical processing to generate cross-sectional representations of the organ.

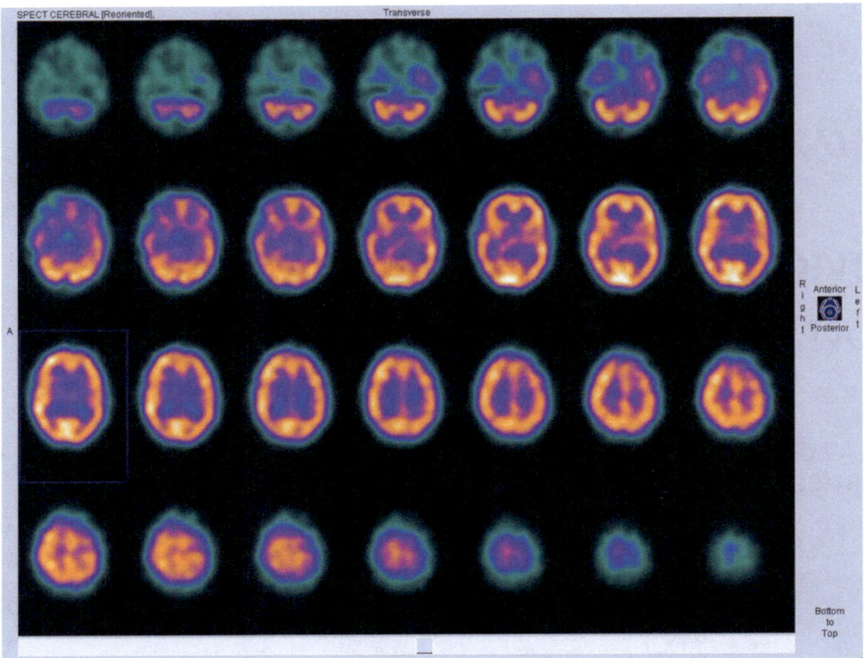

Figure 6.1. *Single-photon emission computed tomography.*

Source: https://media.lex.dk/media/184861/standard_SPECT-billede_normal.jpg, Creative Commons License.

SPECT employs single photons generated by gamma-emitting radionuclides, including ^{99}Tc, ^{111}In, and ^{123}I. This contrasts with positron emission tomography (PET), which employs the paired 511-keV photons resulting from positron annihilation. The basic camera design for SPECT imaging resembles that of a planar camera, augmented by two extra features. The SPECT camera is designed to enable the head to swivel or rotate around the patient to provide different perspectives. It is equipped with a computer that combines the various images to generate cross-sectional views of the organ. To enhance overall efficiency, the majority of SPECT cameras are produced with multiple heads, with two being in the common configuration. The heads are mechanically rotated around the patient to acquire different projection views. Single-headed cameras must complete a full 360° rotation to capture all vital perspectives of the majority of organs. Each head of a double-headed camera requires a rotation of only 180°, while a triple-headed camera necessitates a rotation of merely 120° to get equivalent perspectives (views). The expense of the more heads must be weighed against the advantages of enhanced acquisition speed. Two-headed cameras may possess a fixed, parallel configuration, fixed, or adjustable heads. Fixed, parallel heads or in certain cases opposing heads may be utilized for concurrent anterior and posterior planar imaging or can be rotated as a single unit for SPECT collection. Stationary or fixed, perpendicular heads in an L-shaped configuration are predominantly utilized for cardiac or cerebral SPECT imaging. Adjustable heads facilitate the positioning of the heads in various angle configurations. The adjustable head can be positioned closer to or farther from the other head along the ring, allowing the two heads to be parallel or opposing, perpendicular (L-shaped), or at an intermediate angle. The adjustable dual-headed camera is applicable for planar imaging as well as for both large and small organ tomography. The sequential planar images obtained during tomographic acquisition are referred to as projection views. They are merely intermediate or transitional phases in the process of producing slices. The name cine is employed due to its similarity to films. Tomographic projection images are typically obtained over an arc of 360° or 180°. The camera head 360° rotational arc is commonly employed for the majority of organs. The 180° arc is utilized for organs located on one side of the body, such as the heart. Cardiac views are acquired in a 180° arc ranging from the right anterior oblique position to the left posterior oblique position. The data from this 180° is deemed sufficient, as photons emanating from the right posterior and lateral chest traverse more tissue and experience greater attenuation than those exiting from the left side. 64 or 128 tomographic projections are acquired across a whole 360° arc, while 32 or 64 views are generally taken over a 180° arc.

6.2. IMAGING

For a specified dose of radiopharmaceutical, enhanced images are produced utilizing the count statistics from acquisitions. However, patient comfort and cooperation constrain imaging periods. Standard acquisition times range from 20 to 40 seconds per projection view. The conventional technique for collecting tomographic projection views is referred to as step-and-shoot acquisition. This technique involves acquiring each projection view completely at every angular location i.e., stop. A short interval of several seconds occurs between views to facilitate the automatic rotation of the camera head to the next position. The camera performs a complete rotation around the patient. The camera halts for two seconds after capturing each perspective as the head adjusted for the next view. The cumulative acquisition duration was 1408 seconds. Given that 126 seconds of the total were allocated to pauses, the overall imaging duration amounted to 1282 seconds. In continuous acquisition, data is collected throughout one or several 360° revolutions. There are no interruptions; the rotation is continuous. The camera completes a full 360° rotation every 140.8 seconds. Ten rotations yielded 1408 seconds of image time, in contrast to the 1282 seconds obtained from the step-and-shoot acquisition. Scans on conventional systems were obtained using a circular orbit. The camera head rotated or pivoted at a constant distance from the body center. Due to the circular cross-section, the camera did not approach the organ as closely as possible during a substantial amount of its rotation. The image resolution improves when the camera is positioned as near to the organ as feasible; elliptical and body contouring orbits enable the camera heads to closely adhere to the body contours, hence maintaining proximity to the organ being imaged. The body contouring orbits are facilitated by electronic sensors positioned on the collimators. When the gantry rotates, the detector radii are constantly modified to approach the body while preventing collisions. Extensive patient movement can induce distortions or blurring in a picture. Prior to discharging the patient, it is essential to examine the cine display and/or the sinogram for any indications of patient mobility or motion. Minor patient movement can be rectified by automatic correction algorithms that adjust the projection views to align with the organ of interest. If these algorithms prove unsuccessful, repeat acquisition is advised. A sinogram image consists of a series of slices representing the acquired projection views from 0 degrees to the maximum rotation angle, which can be either 180 degrees or 360 degrees (Fagerstrom, 2020). Every row of the sinogram image contains data collected at varying angles of rotation, however all rows originate from the identical axial (y) point. A distinct sinogram image exists for each slice point along the y-axis (the longitudinal axis) of the patient. A camera engineered exclusively for cardiac imaging, rather than for other organs or the entire body, can achieve a sensitivity advantage by directing comparatively large detectors towards the relatively confined area within the body that houses the heart. Due to the application constraints, designers can enhance the gantry configuration to minimize expenses and elevate patient comfort. Conventionally, cardiac imaging is conducted using a general-purpose dual-headed SPECT camera, with the heads positioned at 90 degrees; the gantry rotates roughly 90 degrees to obtain the 180 degrees of cardiac SPECT data.

6.3. CAMERAS

Cardiac cameras utilize a similar approach to capture photons across a restricted arc while the patient is in an upright or supine position (Slomka et al., 2014). Certain camera designs represent slight modifications of the general camera, wherein many heads articulate to include an arc beyond 180 degrees. Some utilize stationary heads with mobile internal detector assemblies, while others use stationary heads that capture photons through many pinhole collimators. An estimation of radionuclide activity in a lesion can assist in distinguishing benign from malignant lesions and in monitoring disease progression or treatment response. A semi-quantitative assessment of this activity is the standardized uptake value or SUV, which has been utilized in PET imaging for many years. Advancements in SPECT image reconstruction, specifically enhanced attenuation correction maps obtained from CT data and improved algorithms for minimizing scattered photon contributions, have enabled SUV calculations for SPECT data on certain newer scanners. SPECT scanners, similar to PET scanners, inherently quantify counts per second, and translating this to Bq/ml necessitates multiplication by a calibration factor. The calibration factor is frequently referred to as the well-counter calibration factor, as it matches the efficiency factor that quantifies the proportion of counts captured by a well-counter per disintegration of the radionuclide being measured. The calibration factor for each scanner must be established using a phantom experiment in which a known activity is uniformly distributed inside a specified volume (Bq/ml). The scanner subsequently captures and reconstructs images, with regions of interest or ROI positioned at various positions within the uniform distribution, and the counts inside the ROI are averaged. The calibration factor is the ratio of the known activity concentration to the average ROI count value. Positron emission tomography or PET cameras are engineered to identify the dual 511-keV photons produced from the annihilation of a positron and an electron. After emission, a positron traverses a minimal distance before interacting with electrons in the adjacent matter. The paired 511-keV annihilation photons propagate in opposing directions (180°) along a linear trajectory. After acquiring positron emission images, the data are reconstructed similarly to SPECT, except that attenuation correction is consistently applied using either a radionuclide transmission scan or CT data. Positron Emission Tomography (PET) possesses some advantages over Single Photon Emission Computed Tomography (SPECT). The primary advantages are its enhanced sensitivity and resolution, as well as the availability of positron-emitting isotopes for low atomic number elements without adequate gamma emitters. The primary drawback of PET is the increased expense of the apparatus and the less half-life of several of the effective positron emitters. A collimator diminishes sensitivity due to the septa obstructing a portion of the camera crystal surface. A collimator is unnecessary for PET since PET cameras detect the simultaneous and oppositely directed 511-keV photons from positron annihilation to determine the origin of the photons.

6.4. SPECT

In SPECT, the sources of single photons can only be identified if the camera is fitted with a collimator to absorb photons that are not moving perpendicularly to the crystal surface (Ritt et al., 2014). A PET camera, which does not require a collimator, is more sensitive than a SPECT camera and exhibits a larger count rate for equivalent amounts of radioactivity. When two detectors positioned on opposing sides of the annihilation reaction detect concurrent photon hits, the annihilation event occurred along an imaginary line, referred to as the line of response (LOR), connecting these detectors. A coincidence circuit records them as simultaneous or concurrent events. The occurrences resulting from a single positron annihilation subsequent to positron emission are termed true coincidence events. The influence of an unpaired photon, referred to as a singles event, is generally rejected. A single event occurs when an unpaired photon from a non-annihilation gamma ray strikes a detector. A single event is recorded when only one photon from a pair of annihilation photons strikes a detector; the other photon may exit the detecting plane or plane of detection or be absorbed or scattered by the surrounding medium. The photons produced concurrently from distinct locations inside the body may arrive at the crystal simultaneously. These distinct events are erroneously interpreted as coming from the annihilation of a singular positron emission occurring along a trajectory between the two detectors. These photons may originate from annihilation events, wherein one photon from each of two annihilation events is recorded as a simultaneous event, or from the concurrent detection of non-annihilation photons. When the trajectory of a 511-keV photon is modified due to scattering, the site of the annihilation event may be inaccurately inferred to lie along the line linking the detectors. These are termed random events. The likelihood of random events escalates with increased radioactivity in the scanner field of view, rendering them particularly concerning at elevated count rates. Scatter and random events are detrimental as they elevate background counts in images, hence diminishing image contrast. To enhance resolution, certain systems measure time of flight, operating under the premise that the annihilation site can be located along the photon trajectory by recording the arrival time of each photon at the opposing crystals. Unless the event occurs at the precise center of the detection ring, one photon will arrive prior to the other. The time difference will correspond to the difference in distances traversed by the two photons and can be utilized to ascertain the location of the event along the line connecting the detectors.

6.5. DETECTORS

Distinct physiological characteristics of an organ can be visualized; for instance, the brain oxygen consumption and glucose metabolism can be independently observed with $[^{15}O]_2$ and ^{18}F-fluorodeoxyglucose. A PET detector has rings of crystals. The rings may be separated by septa or not. The detector comprises one or many large segmented crystals or a group of tiny crystals. A basic detector unit comprises a collection of small crystals monitored by photomultiplier tubes

(PMTs) (Figure 6.2). One configuration features 36 small crystals organized in a 6 × 6 matrix, connected to four PMTs. Semiconductor materials are replacing PMT tubes. Thallium-doped sodium iodide (NaI(Tl)) crystals were initially employed in PET systems; however, NaI possesses a comparatively low density and is less efficient in attenuating high-energy 511-keV photons.

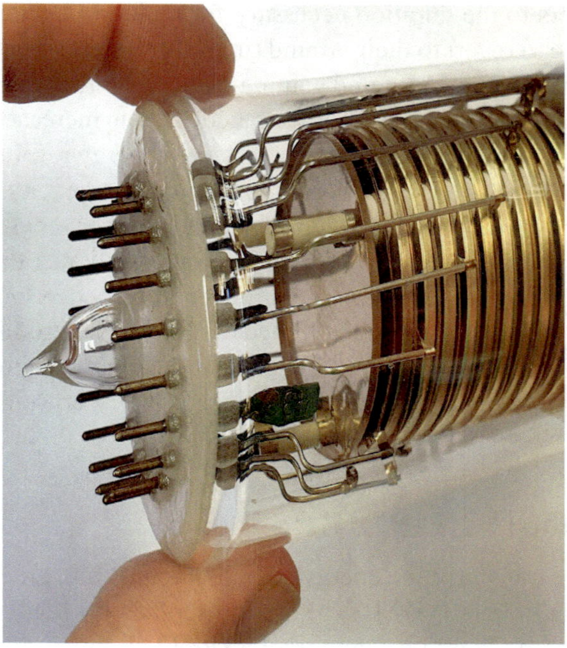

Figure 6.2. *Photomultiplier tubes.*

Source: https://upload.wikimedia.org/wikipedia/commons/a/ac/Technology_photo-multiplier_tube_222_%2852281133769%29.jpg, Creative Commons License.

Thicker crystals were utilized for 511-keV imaging to balance the reduced density, measuring 3.8 cm compared to the 0.6 to 1.2 cm used for detecting lower-energy single photon emissions. Crystals possessing elevated densities and higher atomic numbers (Z), including lutetium orthosilicate (LSO), bismuth germanate oxide (BGO), and gadolinium orthosilicate (GSO), were engineered for 511-keV imaging due to their better stopping power for annihilation photons. The probability of photon interactions per unit volume escalates linearly with density, but the possibility of photoelectric interactions increases sharply with rising atomic number Z, being proportional to Z^3. In detector design, photoelectric interactions are preferred over Compton interactions because the majority of the photon energy is transferred to the photoelectron, which, as a charged particle, will deposit most of this energy in the vicinity of the initial interaction. If the first interaction between a 511 keV photon and the detector crystal is a Compton scattering event, only a small portion of the energy is transferred to an electron and deposited locally, while a significant quantity of the energy from the scattered photon may leave the crystal. LSO and GSO possess a shorter decay time compared to BGO or NaI(TI) for PET imaging. Decay time refers to the duration necessary for radiation-excited atoms within the crystal lattice to revert to their ground (unexcited) state, accompanied by the emission of light photons. A second gamma photon entering the crystal remains undetectable. A higher light yield is advantageous, as an increased photon output per keV of absorbed gamma energy enhances both energy resolution and spatial resolution. Enhanced energy resolution augments the capacity to differentiate between lower energy scattered photons and higher energy annihilation photons. The increased quantity of light photons facilitates the identification of the crystal impacted by the annihilation photon, hence enhancing spatial resolution. Selecting a crystal material necessitates a compromise among cost, density, decay time, and light yield. At present, LSO and a similar composite, LYSO (lutetium–yttrium orthosilicate), are the primary crystals utilized in the production of PET systems (Melcher et al., 2006).

6.6. PHOTOMULTIPLIERS

Photomultipliers transform the energy of light photons into an electrical signal. Silicon Photomultipliers (SiPMs) are swiftly replacing Photomultiplier Tubes (PMTs) in Positron Emission Tomography (PET) scanners due to their small dimensions, better temporal resolution, and enhanced sensitivity (Roncali and Cherry, 2011). They are utilized in PET/MRI systems because of their insensitivity to magnetic fields. They are more sensitive to temperature than PMT tubes. In contrast to single photon imaging systems that utilize many PMT tubes connected to a single large crystal, positron cameras are constructed with many crystal subdivisions monitored by a limited number of SiPMs or PMT tubes. The apertures between crystal subdivisions direct the light photons towards the SiPMs. The localization of the impact site is determined by measuring the light detected by each SiPM; the proximity of the SiPM to the impact site correlates with the intensity of the signal produced by the SiPM. The signals from the SiPM or PMT tubes must be enhanced. The electronic system must identify whether signals originated from paired 511 keV coincident photons resulting from an annihilation event along a line of response between two opposing detectors. This is achieved mainly by quantifying the magnitude or size of each signal, which correlates with the energy of the photon striking on the crystal, and by documenting the time of signal detection. The pulse height analyzer ascertains whether the signal possesses the appropriate amplitude (pulse height) indicative of a 511 keV photon interaction within the crystal. The timing discriminator logs the moment the signal was produced. The coincidence circuit subsequently analyzes signals of sufficient amplitude originating from opposing detectors and ascertains whether the timing of the signals falls within the coincidence time window. The coincidence timing window is generally established between 5 and 15 nanoseconds, depending on the decay period of the crystal material. A coincident event between the detector pairs will be documented by the computer. The paired annihilation photons, when one or both photons have undergone scattering or absorption by adjacent tissue, are typically rejected by the coincident circuit. This occurs due to scattered photons possessing reduced energy and the fact that scattering frequently delays the photon trajectory toward the crystal. Random events are more challenging to distinguish. Coincident events arise when two 511 keV (high-energy) photons, originating from events outside the line of response, reach the opposing detectors within the coincidence time window. Certain systems calculate random event rates by multiplying the coincidence time window with the measured singles rates. Septal rings enhance resolution by diminishing photon scatter from sources external to the plane of a single crystal ring. The scanner sensitivity is diminished due to the rejection of a substantial proportion of actual coincidence events. The elimination of the septa will enhance sensitivity while diminishing resolution. Scans conducted with the septa in position are referred to as two-dimensional scans; scans performed without septa are termed three-dimensional scans.

6.7. SEPTA

The septa obstruct out-of-plane photons, permitting only in-plane or within-plane coincidence events to be documented. The 3-D design enables coincident registration of cross-plane events, where the two 511-keV photons are observed in distinct rings. Septa diminish the frequency of random coincidence events. Positrons traverse a small distance in tissue prior to annihilating with an electron. The camera identifies photons emitted from an annihilation event located at a distance from the actual source of the beta particle emission. Lower energy beta emitters exhibit a limited range, but higher energy beta emitters can traverse a considerable distance before detection. The minimum achievable resolving power of any device for a positron nuclide is constrained by the average range of positrons within the tissue. A further aspect contributing to resolution loss is that 511-keV annihilation photons do not consistently traverse pathways that are precisely 180° apart. The positron-electron combination frequently moves during annihilation, therefore affecting the ejection angle of the 511-keV photons. The detectors, however, presume a conventional 180° emission trajectory of the photons, resulting in an inaccurate localization of the positron emission in these instances. The resolution diminishes towards the periphery of a ring of PET detectors. This occurs because certain photons generated from peripheral annihilation events traverse the detector ring at an oblique angle and may engage with one of multiple detectors along an extended path. When a photon interacts with a detector, it is presumed that the annihilation event occurred along a line of response emanating from the front of the detector, as the depth of contact within the crystal is not documented. This phenomenon is occasionally termed a depth of interaction effect or parallax error (Figure 6.3).

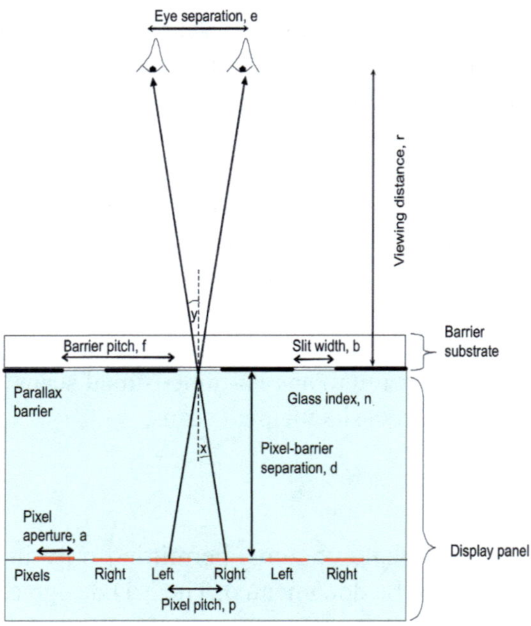

Figure 6.3. *Parallax error.*

Source: https://upload.wikimedia.org/wikipedia/commons/b/b1/ParallaxBarrier-CrossSection.svg, Creative Commons License.

The larger the diameter of the detector ring in relation to the size of the imaged body, the lesser the effect, as the annihilation events will be more centrally positioned and the photons will intersect the detector at a less oblique angle. Due to the comparatively large energy of the positron-annihilation photon and the implementation of coincidence detection, attenuation correction in PET is simpler and precise than in SPECT (Livieratos, 2012). The attenuation coefficient for a 511-keV photon in PET is more consistent across different types of bodily tissues, such as fat, muscle, and bone, compared to the lower energy photons found in SPECT. The total likelihood of absorption for the two photons of the 511-keV pair remains constant, irrespective of the annihilation event position along the line of response. For each position along the line of response, the distance traveled by the photon rises as the annihilation occurs farther from the site of photon emission. This gain is precisely counterbalanced by the reduction in the distance traveled by the photon of the pair in other direction. By summing the length of the path to the right with that of its pair traveling to the left, it becomes evident that the total tissue traversed, and consequently the total probability of absorption for the two photons of a pair, remains constant regardless of the location along the line of response where the annihilation takes place. If one photon of a pair is absorbed by surrounding tissue, the annihilation will not be recorded as a coincident event and will not be included in the count.

6.8. SCANNERS

The attenuation in PET imaging, defined as the reduction of counts resulting from photon absorption prior to reaching the detector, is mathematically corrected using data from transmission scans. Previously, PET scanners employed a high-energy photon source or rotating positron for transmission scans; however, PET/CT scanners utilize CT scan images for this function. Due to the significant difference between X-ray photon energies and annihilation photon energies, scaling factors are required to adjust the linear attenuation coefficients. The quantification of an injected radionuclide absorbed by a tumor or organ is employed in ^{18}F-fluoro-deoxy-glucose (FDG) scans to assist in distinguishing benign from malignant lesions and to monitor tumor development and/or post-treatment regression (Theodoropoulos et al., 2020). The correct measurement of uptake is complicated by various factors, including the inconsistent and frequently poorly understood degree of photon absorption in adjacent tissues. A semi-quantitative metric, the standard uptake value (SUV), is frequently employed as an approximation of the real uptake. The computation of the SUV necessitates a PET estimate of the activity concentration in the tumor or organ in mCi/ml or mBq/ml, the patient's mass in grams, and the administered activity quantity in mCi or mBq. The units of SUV are g/ml; however, as tissue is predominantly composed of water, and one milliliter of water weighs one gram, SUVs are expressed without units. Computed tomography (CT) is a three-dimensional imaging technique utilizing X-ray technology. A CT scanner acquires several planar X-ray images, which are subsequently processed mathematically to provide three-dimensional and cross-sectional images of the body, analogous to the image reconstruction of SPECT scans. Wilhelm Roentgen is recognized for the discovery of X-rays. A contemporary X-ray tube is an evacuated glass or ceramic cylinder featuring a window for the emission of X-rays. In the vacuum tube, electrons are emitted from an electrically heated filament wire, known as the cathode, and are propelled at high velocity towards the positively charged tungsten target, referred to as the anode, due to the large potential difference (or voltage) established between the filament and the target. The majority of these electrons engage with the outer shell electrons of the tungsten target, resulting in the dissipation of their kinetic energy as heat. Approximately 0.2% of the electrons striking the target induce the emission of X-rays through either bremsstrahlung or characteristic radiation.

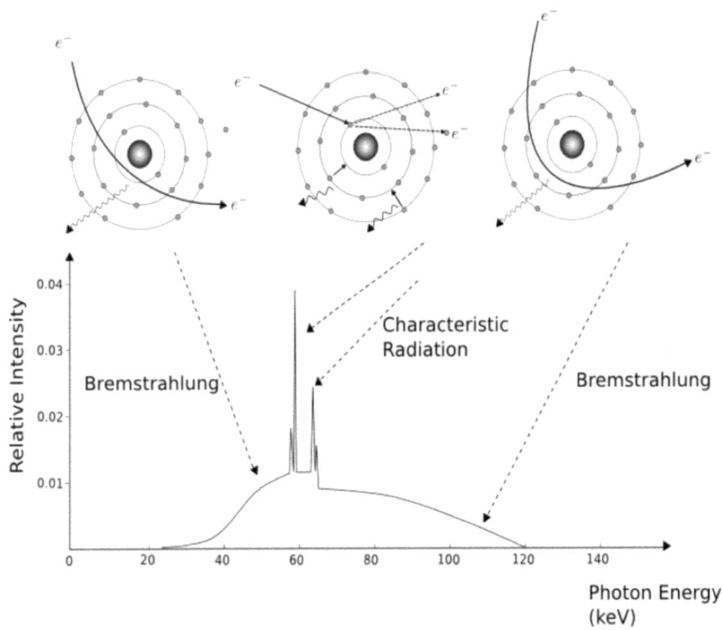

Figure 6.4. *Medical imaging.*

Source: Maier, A., Steidl, S., Christlein, V., & Hornegger, J. (2018). Medical Imaging Systems: An Introductory Guide. Springer., Creative Commons License.

6.9. X-RAYS

Medical imaging (Figure 6.4) employs both characteristic and bremsstrahlung X-rays. Characteristic X-rays possess discrete energy, but bremsstrahlung interactions provide a continuous spectrum of X-ray energies, ranging from 0 keV to a maximum that corresponds to the peak voltage applied between the filament and the target. The energy of each bremsstrahlung X-ray is based on the distance between the traveling electron and the nucleus of a neighboring atom engaged in the bremsstrahlung interaction. The proximity of the electron to the nucleus correlates with an increase in the electron deceleration and the energy of the resultant X-ray from this interaction. The greater the peak voltage applied between the filament and the target in the X-ray tube, the higher the maximum bremsstrahlung X-ray energy and the increased quantity of X-rays produced. The least energetic X-rays in the X-ray beam contribute to the radiation dosage received by the patient, especially to the skin, yet lack sufficient penetration to reach the detectors and enhance image quality. The vacuum tube glass obstructs certain low-energy X-rays, while others are diminished by a filter, often aluminum, positioned in front of the tube window.

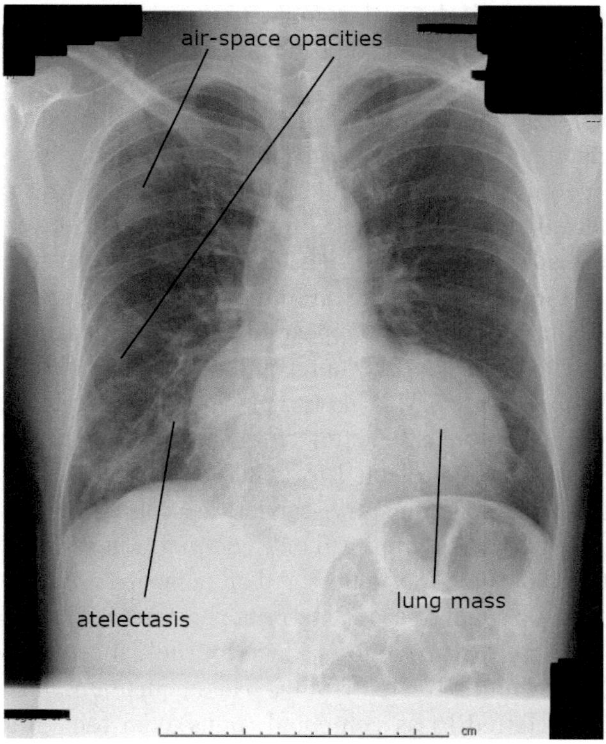

Figure 6.5. *Chest X-ray.*

Source: https://upload.wikimedia.org/wikipedia/commons/6/6c/Chest_X-ray_showing_left_lower_zone_lung_mass_behind_the_heart.jpg, Creative Commons License.

Both the voltage provided between the filament and the target and the attenuation by the filter influence the energy range of the X-rays administered to the patient. The quantity of energy or current utilized to heat the filament for electron emission, along with the duration of this current, influences the total X-ray output. The final quantities are frequently multiplied together and termed mAs (milliamperes-seconds). The X-ray beam is restricted or collimated by lead shutters to prevent radiation exposure to tissues not being scanned (Sharma, 2024). Increased kVP and mAs enhance image quality, but also elevate the radiation dose administered to the patient. When configuring kVp and mAs parameters in CT imaging, it is essential to strike a balance between achieving the requisite image quality for clinical diagnosis and minimizing the radiation dosage to the patient. The bombardment of the object always produces heat. Tungsten is sufficiently dense to halt electrons and can tolerate high temperatures. Heat is drained from the tungsten through a copper backing and fast rotation of the target, which distributes the heat across a wider region and permits a short cooling interval for any specific location before it is re-engaged as the focal point of the electron beam. Additional cooling options encompass reducing the bombardment length and employing external air or sometimes oil cooling. Images produced by projecting X-rays onto the body, commonly referred to as X-rays, are fundamentally shadows. They can be seen as inverse shadows; heavier tissues like bone, which obstruct a significant portion of X-rays, manifest as white on film or digital detectors, whereas lungs, predominantly constituted of air, appear black in the image. An X-ray of a patient utilizing a stationary X-ray source and detector is referred to as a planar image. Chest X-ray (Figure 6.5) is likely a planar X-ray image. If the X-ray data is captured over the complete 360° trajectory surrounding the patient, this data can subsequently be back-projected to generate transaxial slices. The X-ray source and detectors in contemporary scanners are configured accordingly. The X-ray source either rotates within a fixed, complete ring of detectors (rotate-stationary systems) or the X-ray source and an opposing arc of detectors rotate around the patient or synchrony (rotate-rotate systems). The rotate-stationary arrangements are very expensive to produce for multislice detectors, resulting in the prevalence of rotate-rotate systems. The acquisition and reconstruction of X-ray data is referred to as computed (trans)axial tomography (CAT) or computed tomography (CT) scanning (Ter-Pogossian, 1977). The X-ray source is repositioned in increments around the subject. At every position, the X-ray tube is activated, and the patient is subjected to a fan-shaped X-ray beam. The X-rays that are not absorbed by the body are captured by the detectors located on the opposite side of the subject. The detectors consist of ceramic scintillators that, similar to the NaI(TI) crystal, emit light when exposed to X-rays. The scintillator detectors employed in CT scanners must react to the rapidly fluctuating influx of X-rays produced by the CT X-ray source, necessitating an intricate chemical composition of the components in the ceramic compared to the NaI(TI) crystal. These scintillators must have extremely swift decay periods; the initial light production in reaction to X-ray excitation must be quick, and the leftover light within the scintillator, referred to as afterglow, must disappear quickly. Ceramic scintillators are supported by photodiodes that produce electrical pulses or current in reaction to light photons. Photodiodes are semiconductors that operate analogously to photomultiplier tubes (PMTs) by transforming light photon energy into electrical current. Earlier CT scanners included a single row of detectors.

6.10. CT SCANNER

The majority of CT scanners are currently produced with many detector rows positioned adjacently along the z-axis of the scanner, referred to as multidetector or multislice CT scanners (Prokop, 2003).

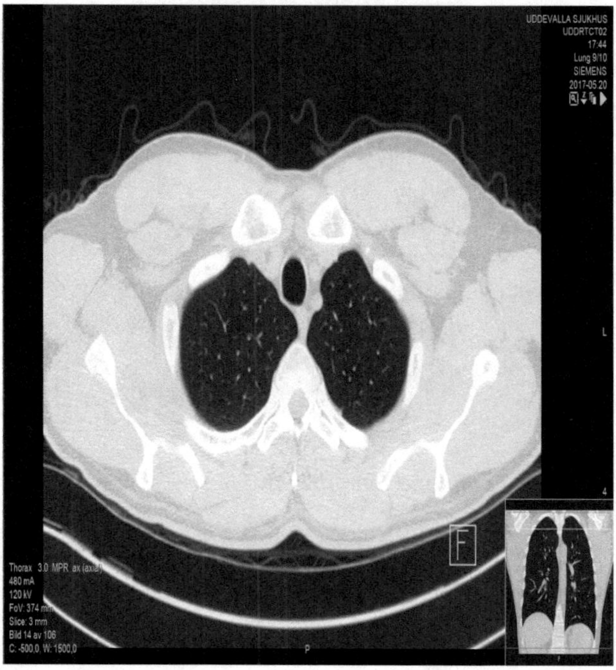

Figure 6.6. *64-slice CT.*

Source: https://upload.wikimedia.org/wikipedia/commons/7/71/High-resolution_computed_tomograph_of_a_normal_thorax%2C_axial_plane_%2814%29.jpg, Creative Commons License.

The use of many detector rows facilitates expedited scanning times by enabling a larger area of the patient to be imaged in a single rotation of the X-ray tube.

The overall quantity of detector rows employed in scanning is dictated by the collimated beam width. Each row contains detectors of uniform dimensions. In contemporary scanners, the innermost rows of detectors comprise smaller detectors than those in the outermost rows. Single detector rows may be utilized to gather a slice of data, or many connecting rows of detectors might be aggregated as a slice. The identification of a scanner, such as 64-slice CT, (Figure 6.6) indicates the quantity of simultaneous data slices, often referred to as data channels, that can be acquired, rather than the total number of detector rows.

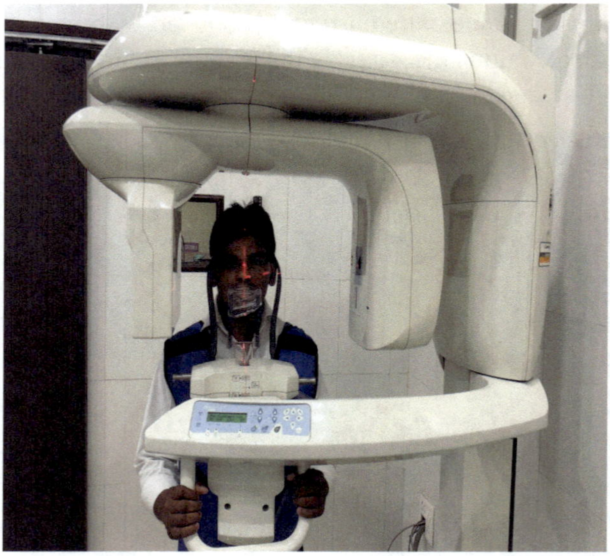

Figure 6.7. *Cone beam computed tomography.*

Source: https://upload.wikimedia.org/wikipedia/commons/8/81/Cone_beam_computed_tomography_%28CBCT%29.jpg, Creative Commons License.

A 4-slice or 16-slice scanner may possess, for example, 16 or 32 detector rows. Assigning the smallest detectors to each slice or channel will yield images with the high resolution; however, this will result in an extended scan acquisition time. Grouping adjacent rows will result in reduced image resolution, but will expedite the acquisition time. Older CT scanners obtained discrete axial slices. The patient bed was advanced in minor increments. Following each increment, the bed was halted, and an axial slice was obtained. Modern CT scanners continually capture data while the bed advances, resulting in an X-ray source trajectory resembling a helix; so, this process is termed a helical scan. The initial advantage is the reduced duration of the study resulting from expedited scanning periods, while the subsequent benefit is the enhanced flexibility in data reconstruction. The operator can select the angle for reconstructing axial slices, from which superior quality coronal and sagittal slices are generated. The helical motion of the gantry is characterized by the rotating speed in revolutions per second and the distance, measured in millimeters, that the patient bed moves along the patient long axis for each gantry revolution. The latter is referred to as the pitch of the helix. Pitch is mathematically defined as the distance traversed by the table divided by the cumulative widths of the detector rows utilized for X-ray data acquisition, which corresponds to the collimated X-ray beam width. Cone beam computed tomography (CBCT) (Figure 6.7) utilizes a source that emits a conical beam of X-rays, which traverses the body to a flat panel detector, and is employed in a restricted range of applications.

6.11. NUCLEAR MEDICINE

Nuclear medicine typically employs older SPECT-CT scanners to acquire nondiagnostic CT scans, which are largely utilized for attenuation correction of the SPECT study (Israel et al., 2019).

Figure 6.8. *Sir Godfrey Hounsfield.*

Source: https://upload.wikimedia.org/wikipedia/commons/0/01/Godfrey_Hounsfield_%281996%29.jpg, Creative Commons License.

They are utilized in angiography to acquire three-dimensional representations of vascular architecture (structure). CBCT is utilized for small animal imaging in nuclear medicine and for three-dimensional imaging in dental and oral medicine applications. The z-dimension, or the length along the body included by the cone beam and flat panel, is typically sufficient to obtain the requisite images in a single rotation of the detector. CT pixel intensities are expressed in CT numbers or Hounsfield units (HU), which are units of attenuation as determined by CT. The Hounsfield unit is named after Sir Godfrey Hounsfield, (Figure 6.8) the developer of the first operational CT scanner, who, along with Allan Cormack, received the Nobel Prize in Medicine in 1979. Magnetic resonance imaging (MRI) is a method that employs a powerful static magnetic field and pulsed radiofrequency electromagnetic radiation to generate a radiofrequency response from the atomic nuclei within the body (Figure 6.9).

Figure 6.9. *Magnetic resonance imaging.*

Source: https://upload.wikimedia.org/wikipedia/commons/8/88/HITACHI%2C_Magnetic_Resonance_Imaging_System%2C_ECHELON_OVAL%2C.jpg, Creative Commons License.

This response provides information regarding the tissue composition. All protons, neutrons, and electrons possess the characteristic of spin. Spin is an inherent characteristic of these particles and is regarded as a type of angular momentum. Momentum is an attribute of a moving object, representing its inclination to maintain motion. Mathematically, momentum is defined as the product of an object mass and velocity (momentum = mass × velocity); so, a huge object moving slowly can possess equivalent momentum to a small object moving rapidly. An object possesses linear momentum when it travels on a linear trajectory, orbital momentum when it revolves around another object, and angular momentum when it rotates about its own axis. Spin is typically depicted as if subatomic particles are rotating in space. Electric fields, denoted as E, engage with electric charges. Magnetic fields, denoted as B, interact with magnetic moments and moving electric charges. Each point in space possesses values of E and B, denoted by vectors. The intensity of the magnetic field is often quantified in Tesla (T). There exists a close link between electricity and magnetism. An electric current is defined as the movement of electric charge.

6.12. ELECTRIC CHARGE

Electric charge in motion generates magnetic fields. An electric current traversing a wire coil generates a magnetic field within the cavity formed by the coil. A rotating magnet generates electric current in conductive materials, a phenomenon known as Faraday induction. Most materials, with the exception of permanent magnets like bar magnets, exhibit detectable magnetism (magnetic moment) only when exposed to an external magnetic field; these materials are classified as possessing inducible magnetic qualities. Subatomic particles like protons, neutrons, and electrons exhibit behavior similar to that of permanent magnets. Similar to spin, magnetism is an inherent characteristic of particles. The spin angular momentum and magnetic moments of subatomic particles are connected. Magnetic resonance imaging was initially termed nuclear magnetic resonance due to its reliance on the interactions between atomic nuclei and various electromagnetic fields generated by several magnet coils (Chizhik et al., 2014). Magnetic resonance imaging utilizes data from the interaction between electromagnetic fields and the nucleus of the hydrogen-1 atom, which consists of a single proton. ^{1}H is utilized due to its abundance in the human body and its net nuclear spin. Imaging of alternative nuclear species, such as P-31 and C-13, is feasible; however, it is not typically conducted in clinical MRI machines. In the absence of an external magnetic field, the magnetic moments of protons (^{1}H) within a sample are oriented randomly in all directions, resulting in an average cancellation of their magnetic moments and a net magnetization of zero. Upon the application of a magnetic field to a sample of protons, the magnetic moments of the protons realign either parallel to the magnetic field or antiparallel to the applied magnetic field. The nuclear magnetic moments possess two potential orientations: aligned with the external field direction or opposed to it. At extremely low temperatures, nearly all the moments will orient in the direction of the external field, resulting in a large total magnetic moment within the material. At body temperature, a greater number of moments are aligned with the external magnetic field compared to those oriented in the opposite direction. At standard temperatures, the aggregate magnetic moments of the proton group yield a minimal net moment, aligned parallel to the external magnetic field, represented along the Z-axis. This quantity is referred to as the longitudinal magnetization (Mz) of the sample. The cumulative magnetization of the protons in the sample is sometimes defined by the net magnetization vector or NMV. The unit of magnetic field strength is the Tesla. The intensity of the Earth magnetic field at the surface is approximately 1/20,000th of a Tesla.

6.13. MRI

Once the magnetic moments (spins) of the protons attain equilibrium, aligning M with the magnetic field, energy in the form of a radiofrequency (RF) wave, sometimes referred to as an excitation pulse, is administered to the tissue. This energy is applied at the Larmor frequency, specifically a limited range of frequencies centered on it, in a direction perpendicular (90°) to the magnetic field (B_0). The application of energy at the Larmor frequency enables the magnetic moments to absorb energy, resulting in their excitation and following alteration of orientation. By selecting the pulse duration, the resultant nuclear magnetic vector (M) is oriented perpendicularly to the Z-axis. MRI is especially effective in distinguishing between normal and pathological soft tissues, which are typically indistinguishable with CT. The repeated application of the RF pulse and echo sampling enables the determination of both the MR signal location and tissue properties. The MR signal waveforms obtained from the body in response to the RF pulse are stored in a matrix known as k-space. Upon completion of the imaging, these waveforms are converted using reverse Fourier transformation into spatial data for visualization. If the magnetic field intensity were uniform throughout the body, all protons would exhibit identical Larmor frequencies, rendering it impossible to differentiate the MR signals originating from various anatomical regions. To address this issue, MR machines are equipped with supplementary coils known as gradient coils, in addition to RF transmitters that can emit RF waves throughout a spectrum of frequencies. Gradient coils are engineered to modulate the background B_0 magnetic field by either augmenting or diminishing it in a linear method along the gradient axis. B_0 is measured in Tesla, whereas gradient coil strengths are typically in the range of tens of milliTesla per meter. The gradient minimally modifies the field strength B_0 at various locations within the body; however, these alterations will manifest as variations in the Larmor frequencies of the protons, thereby affecting the MR signals generated by tissues at these distinct locations. The acquisition of MR data entails the manipulation of gradients, regulation of the transmitted frequency spectrum, and processing of the collected data within a framework known as a sequence or pulse sequence. Many pulse sequences exist on MRI scanners, each with distinct characteristics; yet, they all achieve the fundamental objective of generating a three-dimensional map of several tissue properties, and proton density. A variety of manipulations of radiofrequency pulse applications and variations in tissue response sampling are employed to elucidate additional valuable tissue properties in MR imaging. Hybrid imaging systems were developed to precisely integrate the complementing information obtained from tomographic nuclear medicine imaging, CT, and MRI.

6.14. TOMOGRAPHY

Positron emission tomography (PET) and single photon computed tomography (SPECT) imaging can identify more disease locations for specific clinical diagnosis than traditional anatomical imaging methods like X-ray computed tomography (CT) or magnetic resonance imaging (MRI). The integration of PET or SPECT with CT images obtained sequentially from distinct devices offered a partial resolution to this issue (Bockisch et al., 2009). If patient placement was not meticulously replicated between the investigations, the nuclear medicine and CT images failed to align, resulting in misregistration that caused errors in identifying the anatomical locations of the anomalies observed in the nuclear medicine images. One method for rectifying positioning issues involves the manual or software-assisted fusion of pictures from distinct PET or SPECT cameras and a CT scanner. To enhance the precise registration of PET and CT or SPECT and CT images, the paired imaging scanners are positioned in proximity to one another, utilizing a common bed that traverses both scanners. This configuration enables the patient to retain the same position in both scanners, resulting in significantly improved anatomical registration. The combined machines are referred to as PET-CT scanners and SPECT-CT scanners. This configuration of integrated units also enables the utilization of CT data to adjust nuclear medicine images for the attenuation of gamma radiation as it passes through bodily tissues to the detectors. CT scanners are currently designed as in-line cameras, with a shared or common patient bed or pallet. In PET-CT arrangements, the CT scanner is positioned nearer to the patient. The CT scan is fully obtained before the PET scan is conducted; however, the sequence of the tests may be reversed. All PET-CT scanners are presently produced with multislice CT components. Hybrid SPECT-CT scanners are available in several configurations. Alongside the sequential gantry design, which positions the SPECT camera heads closer to the patient, there exist systems whereby both the SPECT camera heads and the CT X-ray tube and detectors are mounted on a single rotating gantry. The latter approach is more compact, often necessitating less room alterations for installation, and is typically less costly than the twin gantry designs. Besides, due to diminished X-ray tube output, they necessitate reduced room shielding. Single gantry systems have certain restrictions. The integration of the X-ray tube, detectors, and SPECT heads onto a single rotational system restricts the rotation speed, resulting in a slower CT acquisition compared to systems where the CT scanner is housed in a separate gantry. As a result, artifacts from patient motion, both voluntary and involuntary, such as gastric peristalsis, are more prevalent. Due to patient movement and reduced X-ray tube output, the quality of CT images is typically inferior to that of multislice systems. Hybrid cameras have alleviated most of the registration issues between PET or SPECT and CT due to discrepancies in patient placement that arise when patients are physically transferred between separate nuclear medicine systems and CT units. Variations in breathing patterns between CT, which necessitates breath-holding, and PET and SPECT imaging, where breath-holding is unfeasible, can result in misalignment, especially around the diaphragm, leading to misregistration of images in the lower lungs and upper abdomen. Utilizing CT data for attenuation correction may create artifacts into the PET and SPECT pictures in the corresponding regions (McQuaid and Hutton, 2008). To ensure optimal alignment of the diaphragm, several institutions advise patients to breathe normally during CT and PET or SPECT scans. The administration of intravenous and oral contrast in CT imaging enhances anatomical localization; however, the density of the contrast may modify the attenuation maps derived from CT data. The X-ray attenuation will be significantly enhanced at locations with elevated concentrations of contrast, such as the accumulation of oral contrast or vascular filling due to an intravenous bolus. The gamma photon emissions in both PET and SPECT imaging are attenuated by contrast; however, the distribution of the contrast may vary between the acquisition times of the CT and PET or SPECT studies. The CT attenuation map may inadequately represent gamma photon attenuation in these particular regions. The Hounsfield Unit (HU), or degree of X-ray attenuation, will be considerably elevated in the soft tissues where the contrast agent diffuses. The scaling parameters for the attenuation coefficient of soft tissue, derived from noncontrast CT X-ray attenuation, lack precision. Employing attenuation correction data from contrast-enhanced CT investigations to rectify attenuation in PET and SPECT studies may produce artifacts in the resultant pictures. When a contrast CT is required, it is usual practice to initially perform a low-dose CT scan for attenuation correction of the gamma photon images. When the PET or SPECT examination is conducted, contrast medium is injected, followed by the acquisition of a higher quality diagnostic CT scan.

Magnetic resonance (MR) scanners generate images with higher soft tissue contrast and resolution compared to computed tomography (CT) scanners. The integration of this supplementary information with metabolic data from PET scans has demonstrated potential for improved localization and characterization of tissue in certain diseases. Due to the long acquisition time of MR imaging data compared to CT, artifacts resulting from organ motion, particularly in the lungs and bowel, diminish image quality. MRI is more effective for stationary organs, including the brain, limbs, and prostate. Imaging the heart is feasible; however, data capture is synchronized with the cardiac cycle by an ECG connected to the patient. Recent PET-MR scanners have integrated the PET scanner into the MR scanner to facilitate simultaneous image collection. The construction of these integrated components presents significant technical challenges; the intense magnetic fields generated within the MRI tube interfere with the electron flow in traditional vacuum photomultiplier tubes used in PET detectors. Besides, specific materials employed in conventional PET camera design can interfere with the uniformity of the MR magnetic field, resulting in distortion of the MR images. The issues have been addressed by substituting photomultiplier tubes (PMTs) with semiconductor-based photomultipliers, such as silicon photomultipliers (SiPM), (Figure 6.10) which are unaffected by and do not interfere with the MRI magnetic field.

Figure 6.10. *Silicon photomultipliers.*

Source:https://upload.wikimedia.org/wikipedia/commons/1/15/SiPM_IRST.JPG, Creative Commons License.

A significant challenge in PET-MR design has been the generation of precise attenuation maps from MRI data for the purpose of correcting PET picture attenuation. Attenuation correction utilizing CT data is simple, as both the X-rays employed to produce the CT data and the gamma photons captured by gamma and PET cameras are attenuated by electrons within the imaging tissue.

CT images allow for the differentiation of air, water, soft tissue, and bone, and the attenuation coefficients obtained from this data can be utilized to correct SPECT and PET data.

6.15. APPLICATIONS

The signal intensity of MRI imaging is associated with the density and magnetic field-induced behavior of protons in tissue, rather than the electron density of the tissue (Vijayalaxmi et al., 2015). MRI data, thus, cannot be readily utilized to assess the attenuation of gamma rays. On CT, bone, being thick, exhibits the strongest attenuation of gamma photons and X-rays, resulting in a significantly greater attenuation coefficient compared to air, which possesses low density and a low attenuation coefficient. A typical MRI imaging cannot differentiate between bone and air, as both possess a limited number of freely mobile protons, with the majority of such protons in tissue originating from hydrogen nuclei in water molecules. Transmission scans utilizing a germanium-68/gallium-68 to assess tissue attenuation or imported axial CT scans categorized into bone, soft tissue, and air zones were employed for the attenuation correction of PET images obtained from an MRI-PET scanner. Recent methodologies include MRI data for attenuation correction. This technique allows for the segmentation of MRI axial images, similar to CT scans, into regions of air, soft tissue, and bone, which may then be assigned specific attenuation coefficients for the tissue types to correct the PET images. In contemporary nuclear medicine, computers are extensively integrated into the clinical workflow. Software algorithms are essential for image reconstruction from projections and the visualization of digital data on computer monitors. Specialist software may be utilized in data processing for activities such as filtering to diminish specific aspects in images. Reconstruction is the procedure of generating transaxial slices from projection images. In backprojection, the data obtained from the camera are utilized to generate multiple transaxial slices. Backprojection employs two arrays of projection data to reconstruct an image. The array adjacent to the matrix is backprojected across the rows, and these values are incrementally added to the corresponding values of the preceding set. When the counts in each pixel are depicted as dots, a dense core region emerges, which generally aligns with the size and location of the original disk. The extensive bands of dots radiating in four directions from this central density, creating a shape, are artifacts of the backprojection process; they represent residual counts from the backprojection of the arrays. Increasing the number of projection views utilized to generate the image results in residual counts from the backprojections.

CHAPTER 7

RADIATION PROTECTION

Contents

7.1. INTRODUCTION

Various materials positioned between a source and a receptor can influence the quantity of radiation transferred from the source to the receptor. These effects result from the attenuation and absorption of emitted radiation within the source, the encapsulating or enclosing material, or a shielding cover or wall (Figure 7.1).

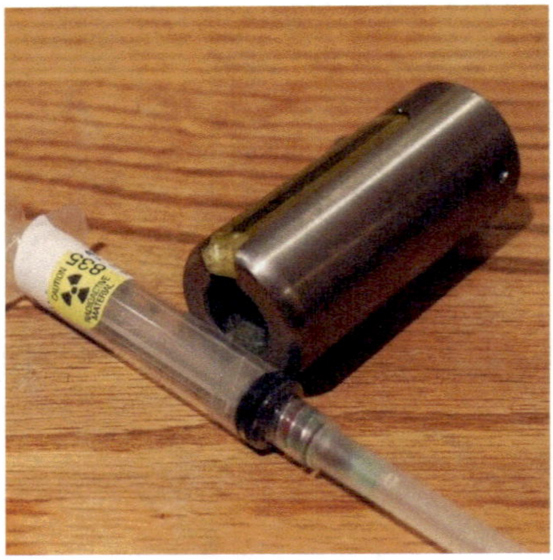

Figure 7.1. *Radiation shield.*

Source: https://upload.wikimedia.org/wikipedia/commons/8/8e/Radioactive_Syringe_and_Shield.jpg, Creative Commons License.

It is important to acknowledge that radiation shielding is a highly intricate or complicated field due to the diverse radiation sources and the many geometric configurations they may present. Alpha particles are easily shielded due to their considerable mass and charge, which restrict their range in diverse materials to tens of micrometers and merely a few centimeters in air (Bawazeer et al., 2023). Only some entities can traverse the 7 mg/cm² dead layer of human skin (epidermis). Within the body, they inflict significant damage. Most shielding problems of alpha emitters are securing the material to prevent it from becoming a contamination source that can be ingested or inhaled as airborne particles. A coat of paint may be employed to diminish severely contaminated regions to non-detectable levels. Ongoing inspections and assessments are necessary to guarantee the integrity of the fixative or paint, particularly in floors. Highly contaminated places can be easily ignored due to insufficient proximity, the utilization of an inappropriate detector, or employing a detector with a thick window. Beta particles possess a restricted range in various media due to their charged nature, resulting in the complete dissipation of their energy through ionization and bremsstrahlung as they move through a medium. The ionization may be extensive due to the small size of electrons. The average range of beta particles does not exceed a few millimeters, indicating that air can substantially attenuate beta particles; the shielding efficacy of denser materials is considerably greater, as the range of beta particles is merely a few millimeters in such substances. This behavior allows for the cessation of all released beta particles by identifying the maximum range of the highest energy beta particles and selecting a medium thickness that meets this range. Most beta shields are constructed using this method, typically with increased thickness. An infinite slab source illustrates several common sources encountered in radiation protection. Such sources can be depicted as a uniform concentration within a slab of material, many of which are sufficiently large to be regarded as infinitely thick and unlimited in size. Volume sources like barrels or tanks containing radioactive material, generate scattered photons as a result of self-absorption by the medium in which they are positioned. Calculating radiation exposure for these cylindrical and spherical geometries is complicated; but, valuable data can be derived by partitioning them into several point sources and aggregating the contributions of each. The usage of various photon absorbers introduces additional complexity due to the distinct effects exhibited by each. The primary consideration is the generation of Compton scattered photons, which can be substantial for low to medium-energy photons in low-Z absorbers, factors that subsequently affect the buildup or accumulation factor.

No exact solutions exist for addressing the difficulties of absorber materials with varying atomic numbers, although some approaches may be employed. A low-Z material like water will yield a larger proportion of scattered photons compared to high-Z materials like lead; so, the low-Z absorber should be positioned nearest to the source, while the high-Z material should be situated externally to capture the Compton scattered photons generated by the initial absorber. If a source is encased in a layer of lead and subsequently submerged in water, the resultant buildup flux would be significantly greater. This is due to the photons entering the water generating a comparatively higher number of scattered photons, attributable to the Compton interaction coefficient in water for low to medium-energy photons. These scattered photons would not be absorbed in the water, as photoelectric absorption is relatively minimal in a low-Z medium. However, if the water shield is positioned next to the source, the scattered photons generated within it will be significantly absorbed by lead through photoelectric interactions, due to lead's high atomic number. While the buildup or accumulation factor is elevated in lead for low to intermediate-energy photons, this is significantly counterbalanced by its enhanced impact on attenuation and absorption. Researchers should prioritize the low-Z substance. If the atomic numbers of the two media differ by no more than 5 to 10, utilize the buildup or accumulation factor of the medium with the greater value, while calculating the overall buildup factor for the combined thicknesses, as the buildup factors do not fluctuate significantly with the atomic number of the medium, except at low energy. If the media differ, with the low-Z medium (first), use the buildup factor of the second medium as though the first medium is absent, as the Compton scattered radiation from the first is of lesser energy and is anticipated to be absorbed by the second. Proton, triton, deuteron, and helium ion beams may exist in the vicinity of accelerators. Being charged particles with significant mass, they will ionize atoms within a shielding material and might be entirely stopped after losing all their energy. The shielding for these particles adheres to the same principles as that for beta and alpha particles, specifically by ascertaining the maximum range of the energetic particles within the beam. Although charged particles may generate some bremsstrahlung, this effect is negligible unless for highly energetic particles, naturally exceeding several tens of MeV (Podgoršak, 2014). Even lightweight materials are useful in housing energetic protons. It is essential to carefully analyze the generation of additional radiations resulting from proton absorption interactions inside the selected material. Shield designs for light ions are based on a penetration thickness for protons of a specific energy. The reference value is subsequently modified to reflect the distinct mass and charge of tritons, deuterons, and helium ions.

7.2. RADIATION SOURCES

Many radiation sources can be effectively modeled as approximate point sources; however, various real-time scenarios cannot be represented in this manner, such as long pipes or tubes approximating a line source, contaminated areas resembling disc planar sources, and diverse volume sources. Diverse practical calculations can be employed to ascertain the photon flux, which can subsequently be utilized to compute radiation exposure. These calculations are conservative, as they often overestimate exposure; but, they get simplified, and the errors in the estimates are minimal. An internal radiation dosage may result from the inhalation or ingestion of radionuclides, direct injection for diagnostic or therapeutic purposes, or dermal absorption. Internal radiation doses cannot be directly measured; they must be computed based on an estimated intake, an estimated quantity within an organ, or the amount excreted from the body. Exposure to an environment containing noble gases like xenon, krypton, or radon may irradiate interior organs; however, exterior exposure typically exists. Submersion exposure is generally regarded as a specific instance of internal exposure. Initially, it was assumed that the deposited radionuclide is uniformly dispersed within the tissue of a source organ. The radionuclide releases energy within a source organ S, which is absorbed by a target tissue T, defined by an absorbed fraction, $AF(T \leftarrow S)$ (Hänscheid et al., 2023). The source organ serves as the target organ; if the deposited radionuclide is an alpha or beta emitter, it will function as the target organ, with all emitted energy being deposited there, i.e., $AF(T \leftarrow S) = 1.0$. For X-rays and gamma rays, $AF(T \leftarrow S)$ typically remains below 1.0 and exhibits significant variation based on photon energy and the masses of the source organ and target tissues. Given that many radionuclides release multiple forms of radiation per transformation, it is essential to consider the energy absorbed from each emitted radiation and its corresponding fraction Y_i of the total transformations or changes that take place. While the instantaneous dose rate holds some significance, the cumulative dose administered over time (total dose) is of relevance.

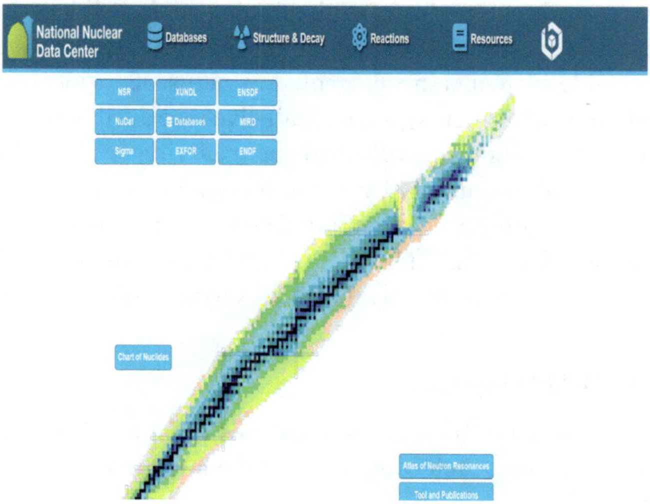

Figure 7.2. *Screenshot of the National Nuclear Data Center website (www.nndc.bnl.gov).*

Source: https://www.nndc.bnl.gov/, Creative Commons License.

All parameters influencing the instantaneous dose rate remain constant for a certain exposure scenario, except for q(t), which fluctuates over time following deposition in an organ due to the cumulative effects of radioactive transformation and biological turnover. Absorbed fractions for photon energies ranging from 0.02 to 4 MeV are recorded for the lungs, liver, and entire body as source organs. Similar data have been assembled by the MIRD Committee for up to 25 source-target organ pairings and can be accessed at www.nndc.bnl.gov (Figure 7.2). Specific absorbed fractions, which incorporate tissue masses based on a reference individual, are readily accessible. The absorption of energy in source organs frequently constitutes the primary factor in the internal dosage from a radionuclide. All energy emitted as particles will be deposited in the source organ; however, only a portion of the photon energy will be absorbed. The dose coefficient (J/kg t) for the source organ can be determined by accounting for the total particle energy absorbed by the source organ mass and incorporating the energy from the fractional absorption of the photon(s). The differential elimination rates of a radionuclide from an organ or tissue typically result from many biological retention compartments, each possessing distinct effective removal half-lives. It is essential to compute the number of transformations occurring in each compartment individually and to aggregate them to ascertain the total dose administered to the tissue. Biological investigations have demonstrated that 10% of cesium deposited in the body is eliminated from a sub-compartment with a biological half-life of 2 days, while the remaining 90% is cleared with a biological half-life of 110 days. Retention functions for stable elements are presented as sums of exponential functions (cesium with two compartments); so, the half-lives are attributable to biological elimination i.e., all cesium, regardless of radioactivity, is cleared in the same manner. It is essential to multiply by $e^{-\lambda rt}$ to incorporate both radioactive decay and biological turnover for each retention/clearance compartment; however, for long-lived radionuclides like ^{137}Cs ($T_{1/2}$ = 30.04 y), the biological half-life for each retention compartment effectively equates to the effective half-life.

7.3. RADIATION DOSE

The quantity of changes resulting from the deposition of an initial activity q0 is determined by integrating each component of the retention function throughout the specified duration. The inhalation or ingestion of radionuclides complicates internal dose assessments, as these substances must traverse the respiratory system and/or the gastrointestinal tract prior to entering the bloodstream, from where they are subsequently absorbed by organs.It is essential to ascertain radiation doses to the lungs and the segments of the gastrointestinal system as the radioactive material traverses these areas, in addition to doses to tissues where it subsequently accumulates in the surrounding tissues (Figure 7.3). The internal radiation dose is directly correlated with consumption; hence, restrictions for internal radiation dose are built on establishing an annual limit of intake (ALI) for each specific radionuclide. These restrictions, which have fluctuated throughout time, are currently based on an assessment of risk. Prior guidelines for internal dose, established during the 1940s, were mostly based on risk; however, this was not explicitly articulated, particularly not in numerical terms.

Risk has consequently emerged as the primary or principal protection standard, whereas dose, whether administered internally, externally, or both, serves as a secondary standard for fulfilling the risk standard. Before 1977, internal dose limits were established for a critical or key organ, which, based on its sensitivity, was restricted to 15 rad/y for most tissues, a factor of three higher than the total body radiation limit. The radiation limitations for specialized tissues, including the thyroid, skin, and gonads, were established at 30, 30, and 5 rad/y, respectively, whereas skeletal tissue for bone investigators was permitted up to 50 rad/y. Permissible dose limits for particular tissues were permitted to exceed those for the entire body due to their comparatively reduced risk to the overall organism. The organ doses were evaluated independently from external doses based on the assumption that one would predominate in an occupational context or setting.

Figure 7.3. *Radiation: Gastrointestinal system.*

Source: https://upload.wikimedia.org/wikipedia/commons/3/32/Dr_nilay_shinde_md_radiologist_working_on_mri.jpg, Creative Commons License.

In 1977, the ICRP calculated that a whole-body dose of 1 rem or 0.01 Sv to 10^6 exposed individuals would result in a risk of 165 significant health consequences. These effects were observed in six primary tissue types, together with additional tissues, resulting in a total body risk considered tolerable if the total effective dose equivalent (TEDE) for each individual did not exceed 0.05 Sv (5 rem) annually. The ICRP framework further restricted the 50-year committed dosage for any tissue to 50 rem (0.05 Sv) each year to prevent non-stochastic consequences, including fibrous tissue damage. Both stochastic and non-stochastic dosage limitations assume the absence of external or additional doses. In 1990, the ICRP re-evaluated the risk data and determined that a linear non-threshold model accurately represented risk compared to the linear quadratic model employed in its 1977 assessment (Stewart et al., 2012). The 1990 recommendations have two primary effects: they restrict the average yearly dose to 2 rem and confirm the prior limit of 0.05 Sv (5 rem) per year, while imposing a restriction that the 5-year average must not exceed 0.1 Sv (10 rem). Radiation protection rules in the USA are currently based on a risk assessment aligned with Federal Radiation Guidance endorsed by the EPA and ratified by the president in 1987. Given that the basic protection requirements were established using risk data pertaining to whole body doses, a mechanism is required to modify doses to internal organs to achieve an effective whole-body dose. These doses yield differing levels of risk to the individual; so, the overall risk is ascertained by assigning a relative weighting factor w_T to each of the vulnerable tissues, and the cumulative risk is derived by summing these to achieve a dose equivalent to that of the uniformly irradiated total body. The models have been established for determining doses from the inhalation and ingestion of radionuclides, specifically a pulmonary deposition/clearance model and a gastrointestinal tract model.

Tissue dose is determined by the chemical and physical properties of the radionuclide. It is essential to ascertain radiation doses to the lung and different segments of the gastrointestinal system when the radioactive material traverses these areas, in addition to doses to tissues where it eventually accumulates. A specialized model has been created for bone dosimetry due to the different distribution and deposition of bone-seeking radionuclides. Inhaled radionuclides are initially deposited in different lung compartments and then cleared to other tissues either through direct absorption into the bloodstream or by clearance to the gastrointestinal tract. Radioactive material transferred from the lung to the gastrointestinal system is eliminated in the same way as ingested material; hence, interconnected pathways. The gastrointestinal tract comprises four segments: the stomach (ST), the small intestine (SI), and the upper (ULI) and lower (LLI) large intestines. A fifth compartment is designated for fecal removal and is considered a holding compartment, so it is termed a bucket compartment.

7.4. RADIONUCLIDE EFFECT

A radionuclide entering the stomach can deliver a radiation dose to the stomach wall as it transitions to the small intestine, where it subsequently imparts a dose to the small intestine wall. Based on its solubility, the radionuclide may be assimilated into the bloodstream and migrate to various organs and tissues, thereby delivering a dose to them. The residual segment proceeds to the upper and lower large intestine, where it administers a dose to the walls of the ULI and LLI as the substance is transported for final elimination. The gastrointestinal system serves as a significant pathway for the translocation of both inhaled and ingested radioactive substances to various tissues and organs throughout the body. Two models, created by the ICRP Task Group on Lung Dynamics, are employed to elucidate the deposition and removal of radionuclides in the lung. The models serve as the foundation for ICRP calculations of internal dose for inhaled radionuclides.

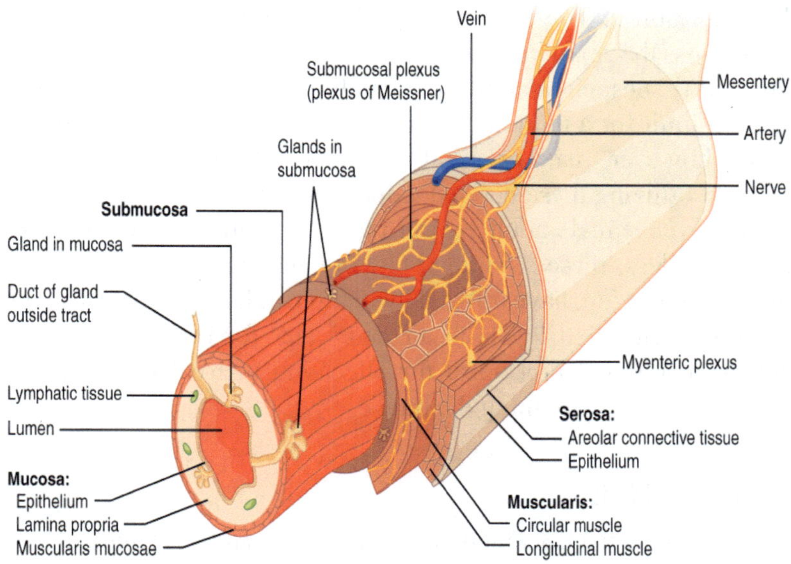

Figure 7.4. *Gastrointestinal tract.*

Source: https://upload.wikimedia.org/wikipedia/commons/b/b1/2402_Layers_of_the_Gastrointestinal_Tract.jpg, Creative Commons License.

The lung model was upgraded in 1996 but has not been integrated into US laws governing internal radiation dosage, despite its utilization by ICRP in the computation of updated dose factors for inhaled radionuclides. The ICRP-30 respiratory model categorizes the deposition of inhaled radioactive substances into four primary regions: the nasal-pharyngeal or NP region, the trachea and bronchial tree or TB, the pulmonary parenchyma or P, and the lymphatic or L system, which is viewed as a distinct tissue compartment linked to the pulmonary region. Radioactive material that enters the systemic blood supply may be absorbed in varying quantities by systemic organs and tissues, circulate within intercellular fluid, or be removed directly. The temporal rate of translocation and retention in the lung, blood, and many interrelated tissues can be represented as a system of first-order kinetics using a series of differential equations. The 1996 model (ICRP) incorporates additional regions to address complications identified since 1966; it also categorizes the clearance of radionuclides into three classifications similar to those in the ICRP 30 model: fast (F), medium (M), and slow (S) (Khalil, 2021). A dynamic gastrointestinal tract model, based on linear first-order kinetics, has been employed to ascertain the translocation and retention of radioactive substances deposited in the stomach (ST), which subsequently traverse the small intestine (SI) and the upper (ULI) and lower

(LLI) large intestines. The gastrointestinal tract (Figure 7.4) model involves absorption across the small intestine wall into the bloodstream, allowing uptake by organs and tissues, circulation in intercellular fluids, or direct excretion. A concentration limit for a radionuclide in water may be suitable for the general population; however, national and international standards organizations advise against utilizing these derived values for the general public, as they are based on a reference male adult worker rather than an average member of the public. Nevertheless, these warnings, few organizations, like the US Nuclear Regulatory Commission, have utilized inhalation and ingestion ALIs, modified, to establish concentration levels in air and water for the overall population. An efficient operational procedure for allocating internal radiation doses to individuals involves ascertaining the intake.

7.5. EXPOSURE

Individuals enveloped in a cloud of radioactive material (submersion) may experience a type of internal radiation exposure; but, in most instances, they receive complete body external exposure due to the concentration of radionuclides in the surrounding air. Noble gases and tritium are the primary radionuclides during immersion exposure, with air concentration serving as the limiting factor to prevent surpassing the TEDE limit. The effective whole-body dose typically serves as the limiting factor; however, non-stochastic effects on the skin or lens of the eye may dominate in certain instances. When the limiting concentration is determined by non-stochastic effects, the specific tissue is also identified. Individuals may be exposed to tritium either as elemental ^3H or by the consumption of tritiated water, known as HTO or tritium oxide, ^3H_2O. Exposure to elemental tritium results in a negligible accumulation in body water from inhalation, with low absorption through the skin; the residue, which is challenging to quantify, will be exhaled, where it may undergo oxidation and become available for further exposure. Tritium exhibits fast changes in form, rendering it a confusing radionuclide due to this variability. Most radiation protection assessments for tritium are based on its oxide form, as it is directly assimilated into body water, resulting in whole-body exposure. In an individual exposed to airborne tritium oxide, it is assumed that 50% of the inhaled intake is also absorbed dermally; hence, an individual inhaling 200 Bq of ^3H_2O in a tritium environment will additionally absorb 100 Bq through the skin, resulting in a total body water consumption of 300 Bq (Stabin, 2003). Tritium entering the body disperses swiftly in 42 liters of body fluids, where it is evenly distributed inside and irradiates 63 kg of soft tissue. The behavior of emitted radionuclides in environmental pathways is determined by empirical models that consider their transport and dilution in air, groundwater, surface water, and terrestrial ecosystems. These models yield radioactivity concentrations that can subsequently be utilized to ascertain intakes and radiation doses to individuals. Such intakes are frequently regulated by a complicated interconnected chain like downwind air concentrations generated by a source, subsequently leading to human exposure through diverse patterns. The concentration of a radionuclide in an environmental medium is dictated by: a source term measured in units of activity per unit of time, a dispersion term, a velocity term indicating the rate of input of the diluting medium, and, an expression that delineates the area through which the released material traverses.

7.6. DISPERSION

The concentration of a radionuclide released into an environmental medium is directly proportional to the release rate; inversely proportional to the flow velocity of the mixing or dispersion medium; and inversely proportional to the area through which the contaminant-diluent mixture flows. Surface water dispersion can be depicted as the fusion of released radioactivity into a liquid flow contained within a conduit, channel, or a naturally occurring stream. An example involves a specified quantity of radioactivity in a holding tank that is released at a consistent rate into flowing water over an extended duration. Under these conditions, the material will create a mixing zone of heterogeneous concentration near the discharge point, but it is expected to achieve uniform mixing a short distance downstream. The concentration beyond the mixing zone is influenced by the input rate, stream flow rate, and the stream width and depth subsequent to exiting the mixing zone. Atmospheric emissions of a pollutant are typically presumed to originate from a source that can be accurately represented as a point emitting or discharging material at a constant rate. The released material enters an atmospheric volume characterized by a specific turbulence profile, flowing past the release point at a velocity μ (m/s), resulting in its subsequent mixing. In the presence of uniform turbulence within the atmospheric volume adjacent to the release point, the contaminant discharged within one second will, under specific conditions, be diluted in an atmospheric volume equivalent to the product of width y (m) and height z (m) multiplied by the downwind air translocation rate, denoted as u (m/s). Lapse conditions arise when the air temperature gradient exceeds the dry adiabatic lapse rate. When a parcel of air contaminated mixture in a plume is rapidly forced upward, its absolute temperature will fall due to adiabatic cooling; but, its temperature relative to the surrounding atmosphere will be higher, resulting in increased buoyancy. If the parcel descends, its temperature will decrease more swiftly than that of the surrounding atmosphere, and its comparatively higher density will lead it to descend further. Intense lapse conditions generate looping plumes that extend over a wider area; hence, they are intrinsically unstable and significantly conducive to pollutant dispersion. Temperature inversions produce relatively thin plumes, due to relatively stable conditions. Upward displacements of the parcel result in adiabatic cooling, rendering it denser than the surrounding atmosphere; hence, its ascent ceases and it reverts to its initial altitude. Any descent of the parcel results in a temperature elevation compared to the surrounding atmosphere, and its buoyancy causes it to ascend back to its initial altitude. The atmosphere in this instance is characterized as steady, with limited pollutant dispersion. This situation also arises for releases into an isothermal profile, but to a lesser extent. Light winds typically correlate with stable conditions; when the wind direction remains constant, the plume forms a ribbon extending downwind at the stack altitude. If the wind direction varies significantly, the plume disperses horizontally like a fan, hence the term fanning. A fanning plume is not detrimental to the dispersion of effluents, as the plume is sufficiently wider and does not make contact with the ground.

7.7. LOFTING

Lofting is a phenomenon that occurs when a temperature inversion is present under the plume elevation (Weil, 1988). It generally occurs shortly after sunset when the earth radiative cooling establishes a nighttime temperature inversion, resulting in stable air under the plume while unstable or neutral conditions prevail above it. Lofting is an advantageous phenomenon as effluents can disperse vertically while being prevented from reaching the ground by the stable air underneath; hence, dispersion occurs over extensive distances inside huge volumes of air. The Gaussian plume model (Figure 7.5) is applicable for stack emissions when the release resembles a point source and the wind field exhibits homogeneous turbulence. These conditions apply to tall and short stacks positioned to minimize turbulence from adjacent structures and terrain. A general guideline is that the stack height should be 2.5 times greater than any nearby obstructions that could generate turbulence, leading to inhomogeneities at the release point. Gaussian plume conditions can arise for shorter stacks under moderate to low winds, but may not be present for taller stacks in high wind speeds. The stack influences plume dispersion, with certain effects facilitating dispersion and others inhibiting it. The effluent temperature can provide buoyancy, while its momentum can influence the altitude at which the plume stabilizes. These impacts are being evaluated for effective stack concentration effects, stack height, and stack turbulence effects. The effective stack height is the altitude above the actual stack height at which the plume stabilizes, commonly known as plume rise. The extent of plume rise is determined by the efflux velocity that imparts momentum to the effluent upon exiting the stack, the horizontal wind speed, and the temperature differential between the effluent and the ambient air stream.

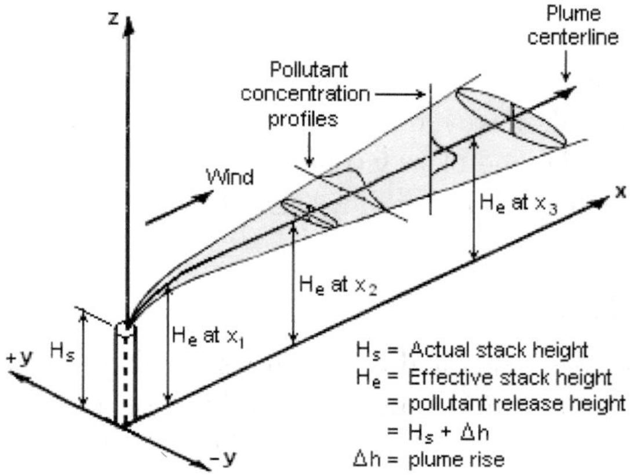

Figure 7.5. *Gaussian plume model.*

Source: https://upload.wikimedia.org/wikipedia/commons/1/10/Gaussian_Plume.png, Creative Commons License.

A heated plume will exhibit buoyancy and, if released at a high efflux velocity, may ascend above the physical stack height before being stabilized by a combination of wind speed and the cooling influence of the ambient air. Plume concentrations may be significantly elevated during temperature inversions, which generally occur in conditions of minimal wind and clear night skies. These conditions facilitate radiative cooling of the Earth surface, resulting in the near-ground air temperature becoming comparatively lower than the layers above, so establishing a zone of considerable atmospheric stability. As the sun rises, it heats (warms) the earth, generating a zone of turbulent air that first forms near the surface but ascends due to thermal eddies. As the turbulent air zone rises to the plume altitude, the contained plume material is swiftly disseminated to the ground, resulting in comparatively elevated concentrations at ground level. Effluents released during this interval are trapped by the overhead inversion and will be disseminated to the ground by turbulence in the fresh heated unstable air. The dispersion of fumigation for an elevated receptor can be modeled by modifying the He value in the fumigation equation, to accommodate the elevated point or zone of interest. This effectively reduces the area through which the plume material flows at the receptor site, thereby elevating the ground-level concentration due to the trapped plume. This conservative method is adopted for protection, as the vertical dispersion depth is expected to be greater due to the elevated position of the receptor, particularly if situated on a small hill or as a result of a progressive terrain modification. The Gaussian diffusion equations were formulated for ideal scenarios involving point source emissions over relatively flat, undisturbed landscapes, when atmospheric turbulence, as defined by the vertical temperature profile, is homogeneous. Such idealized conditions are rarely encountered; however, in actual applications, point source equations may be utilized for flat and clear terrain, as well as for release points that are 2.5 times higher than adjacent buildings or terrain anomalies (rock outcrops).

7.8. TURBULENCE

Mechanical turbulence generated by buildings or other structures can entirely govern the dispersion pattern due to a combination of low efflux velocity, moderate to strong winds, and relatively short release points. A prevalent scenario involves a small stack atop a large building. Effluent discharged under these conditions will be incorporated into the building wake, necessitating specific considerations for assessing plume concentrations. A virtual source method may also be employed, as mechanical turbulence generated by the obstacle allows the released contaminant to traverse along the ground. The plume concentration can be modeled as a ground-level release by positioning a virtual source at an upwind location, ensuring that the plume generated by this source has a horizontal dispersion that completely involves the building as it traverses past it. Contaminants may be discharged in a single event during situations due to over pressurization, potentially resulting in an explosion, or from a sudden leakage.

These releases are defined as short emissions in which a specific volume of material disperses downwind, expanding due to atmospheric influences. One or more of these locations may indeed exert control, and if it is feasible to restrict releases to specific intervals, the impacts of the releases can be optimized for the meteorological conditions, particularly if real-time systems and computational categorization are accessible. It is standard procedure in the siting and licensing of a nuclear facility to utilize meteorological data from local weather stations to assess the potential impact of projected releases. The data are subsequently augmented with many years of on-site meteorological data, collected before and throughout operations, using towers designed to detect wind direction, speed, and vertical temperature profiles. Many facilities lack data for the relevant periods, necessitating the retrospective construction of such information, usually derived from local weather stations at airports or adjacent urban areas. This information may be sufficient if the data is representative of the site conditions; however, it represents a compromise between the absence of data and the preferable on-site measurements.

7.9. PLUMES

As plumes travel downwind, they encounter many factors, like dispersion and dilution, that typically diminish their concentration (Yee and Biltoft, 2004). Large particles may precipitate, and plume contaminants may be attached to vegetation, land surfaces, or other entities as a result of physical and chemical mechanisms. The impact of these processes on plume concentrations is relatively minor compared to emissions from the source, and they are typically disregarded in calculations of downwind concentrations. Their primary consequence is the deposition of pollutants on vegetation and the ground, where they may enter food chains and/or serve as a source of direct exposure from the surface. Deposition velocities are influenced by wind velocity, as the vertical concentration profile varies with it. The extent of deposition is significantly affected by sorption properties, surface roughness, and the chemical composition of the effluent. Even the accurately recorded deposition velocity data exhibit an uncertainty that may fluctuate greater. So, these values are constrained and should be applied carefully considering specific conditions, such as the physical and chemical states of the radionuclide, the deposition medium, and atmospheric factors like stability and wind speed, all of which can significantly affect deposition. Persistent pollutants that accumulate over time may get resuspended, resulting in increased concentrations in the surrounding environment or perhaps being translocated to different locations.

The resuspension of deposited material that remains near the surface is primarily influenced by wind velocity, surface roughness, and other mechanisms that expand the contaminated region. Attention has been directed towards land areas polluted with plutonium due to accidents such as leakage from stored drums at the Rocky Flats Plant in Colorado and partial detonation of nuclear weapons in Spain. A similar situation involves redistributed uranium tailings that may be reintroduced into the atmosphere. Both insoluble substances that persist near the surface unless covered in some manner. The airborne concentration of resuspended long-lived materials is directly correlated with the aerial contamination of the soil and its distribution depth. Surface-level pollution is particularly significant, especially in loose, dry soils inside open areas exposed to heavy winds. Contaminants may be eliminated from a plume through precipitation, leading to the wet deposition of plume particles onto vegetation, the ground, and other exposed surfaces. Wet deposition and dry deposition both contribute to aerial pollution; however, they need to be evaluated independently, considering the possibility of their simultaneous occurrence during a precipitation event. The extent of washout depends on the size and dispersion of raindrops as well as the physicochemical properties of the plume. The parameters, which are functions of the spatial coordinates, render the degree of washout a spatially dependent variable; but, due to limited empirical evidence, it is practically considered to be constant throughout space.

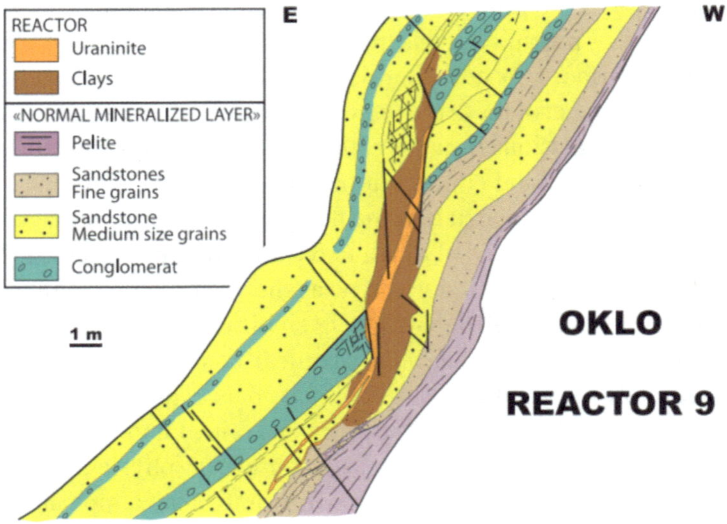

Figure 7.6. *Natural reactor (Gabon).*

Source: Bentridi, S.-E., Gall, B., Gauthier-Lafaye, F., Seghour, A., & Medjadi, D. E. (2011). Inception and evolution of Oklo natural nuclear reactors. Comptes Rendus Geoscience, 343(11), 738–748. https://doi.org/10.1016/j.crte.2011.09.008, Creative Commons License.

7.10. NATURAL REACTOR

A criticality event of a natural reactor occurred in Gabon, Africa, (Figure 7.6) approximately 2 billion years ago (Gauthier-Lafaye et al., 1996). Analyses of uranium ore from the location revealed that the ^{235}U level was below the 0.72% typically seen in contemporary natural uranium, indicating that the ^{235}U had been depleted due to fission. The Gabon natural reactor demonstrates the interaction of several conditions crucial for the occurrence of nuclear criticality. The ore body was large to ensure a critical mass. Under arid conditions, the surface layer would crack, allowing rainwater to accumulate in the crevices and create a moderator zone to decelerate neutrons to thermal energy, facilitating fission of ^{235}U in the ore. The ^{235}U concentration was adequate to surpass neutron absorption losses in hydrogen atoms found in light water. Spontaneous fission of uranium would generate neutrons to initiate the reaction. The heat of reaction would vaporize the water, ejecting it from the matrix and halting the reaction, which would resume upon the accumulation of more water in the cracks, triggering a spontaneous fission event to initiate a new chain reaction. This sequence was repeated many times over hundreds to thousands of years until the semi-ideal conditions altered, potentially due to the reconfiguration of the ore body from the energy of the reactions, climatic shifts, and/or a decrease in the abundance of ^{235}U below the threshold required to maintain a chain reaction with a light water moderator. In 1942, Enrico Fermi accomplished the first human-induced sustained chain reaction, replicating conditions that occurred naturally almost 2 billion years ago. Due to the significantly reduced ^{235}U content in uranium, Fermi opted to incorporate natural uranium metal within a graphite moderator in an arrangement sufficient to sustain the reaction once initiated.

7.11. THREE MILE ISLAND (TMI) ACCIDENT

In 1979, a crucial incident in the history of nuclear energy occurred at the TMI Unit 2 reactor in Pennsylvania (Figure 7.7) (Filburn and Bullard, 2016a). The reactor was functioning at approximately its maximum power output of 2700 MWt. The situation had reached a critical state for the first time precisely one year prior to the catastrophe. The coolant temperature was 600 °F and the system pressure was 2,200 psig, both of which are standard settings for a pressurized water reactor. The incident was not a criticality event; it contained overheated fuel, resulting in the release of fission products within the reactor, some of which escaped into the environment. The incident was essential as it highlighted design deficiencies and illustrated that operator training, procedures, and communication are crucial components to mitigate human frailty. At approximately 4:00 AM on March 28, a condensate pump in the secondary loop stopped operation, which is not uncommon. At that moment, there was a disruption in the flow to the feedwater pumps responsible for circulating water to the steam generators.

In the absence of suction, they also ceased operation, as intended. This resulted in an automated shutdown of the turbine; without feedwater, steam cannot be generated, leading to the turbine shutdown. Pressurized water reactors (PWRs) are equipped with auxiliary feedwater pumps to provide feedwater to the steam generators during such incidents. The systems activated automatically to provide water to the steam generators; however, the valves had been left closed by the maintenance crew, a situation that may be deemed the initial human error in the incident. In the absence of coolant water, the secondary side started boiling dry. Eight minutes passed before the closed valves were identified and opened; however, by that point, the steam generators had run empty, and complications followed. The reactor safety systems, if properly engaged, could have rectified the problem; instead, other complications, many initiated by the operators, followed. In the absence of water in the secondary system, the reactor heat could not be dissipated, resulting in an increase in pressure within the primary loop as the reactor continued to operate. This resulted in the activation of the pressure relief valve atop the pressurizer.

Figure 7.7. *Three Mile Island (TMI) Unit 2 reactor.*

Source: https://upload.wikimedia.org/wikipedia/commons/a/ac/TMI_cleanup.png, Creative Commons License.

This lets primary coolant water exit the valve and enter a discharge tank designated for such events. However, given the absence of feedwater to the steam generators for heat removal from the primary loop, the pressure began to rise again, resulting in the automatic shutdown of the reactor. A malfunction is present; but, the reactor remains secure. The pressure relief valve remained in the open position. Had the valve closed, the system pressure would have stabilized, the automated shutdown system would have normalized, and the operator could have initiated a shutdown, allowing for the restoration of power after the feedwater pumps and other components returned to normal operation. However, with the pressure valve jammed in the open position, the system pressure activated the emergency core cooling system (ECCS), to dissipate the fission product decay heat generated by the fuel. A human error occurred when the operators failed to detect the blocked valve, potentially lacking telemetry to indicate its status. The operators deactivated the ECCS, which was a significant error. The primary system pressure decreased further with the ECCS deactivated, leading to the formation of steam bubbles, which subsequently induced a vibration in the coolant pumps. The operators worsened the issue by shutting off the pumps, believing that cooling could be achieved through natural convection. The steam voids obstructed convection movement, resulting in the core overheating as all methods for dissipating the fission product decay heat were rendered inoperative. The water level decreased dramatically, exposing the fuel. In the absence of cooling, the fuel cladding started to deteriorate, resulting in the emission of fission products. Simultaneously, water started reacting with the overheated metal, resulting in a hazardous metal-water reaction that released hydrogen. This resulted in an explosion; however, it was effectively contained by the containment structure. The coolant containing fission products continued leaking from the jammed pressure relief valve, initially flooding the discharge tank and subsequently spilling onto the containment floor. The sump pump activated, discharging contaminated water to the auxiliary building outside the containment structure, allowing volatile fission products to leak into the atmosphere. This pathway emitted approximately 10–20 Ci of ^{131}I and around 10^6 Ci of ^{133}Xe. The maximum dosage to the public was assessed at 80 mrem for a person presumed to remain outside for 11 days. The remaining fission products, including the majority of the ^{131}I, were retained in the primary coolant water within the containment structure. The TMI incident resulted from operator error, worsened by design defects and equipment malfunction, particularly the blocked pressurizer valve. Operator errors were prevalent, including the closure of valves in the feedwater line to the steam generators, misinterpretation of the pressurizer status, and the deactivation of both the emergency core cooling pumps and the reactor cooling pumps. The TMI incident prompted substantial measures that have enhanced the safety and efficacy of all nuclear generating plants. The measures are an increase of qualified operational staff, enhancement of training and licensing protocols for operators, evaluations of control room design to incorporate human factors, the introduction of advanced detectors to ensure operators are continuously informed of reactor status, hydrogen detection apparatus, advancements in the monitoring of emergency conditions, enhanced communication between regulators and facilities; and improved emergency readiness. The nuclear industry collectively undertook these efforts mainly through the establishment of the Institute of Nuclear Power Operations, were senior management exchange experiences and information.

7.12. CHERNOBYL NUCLEAR POWER PLANT

A significant reactor accident emerged at Unit 4 of the Chernobyl nuclear power plant near Kiev, Ukraine, in 1986 (Babenko et al., 2007) (Figure 7.8). The Chernobyl-4 reactor was an RBMK design including a graphite moderator interspersed with more than 1600 pressure-tube fuel channels, each spaced 25 cm apart within the graphite. Clusters of 18 enriched (2% ^{235}U) uranium oxide fuel assemblies were positioned within each pressure tube, and the fission heat was dissipated using light water.

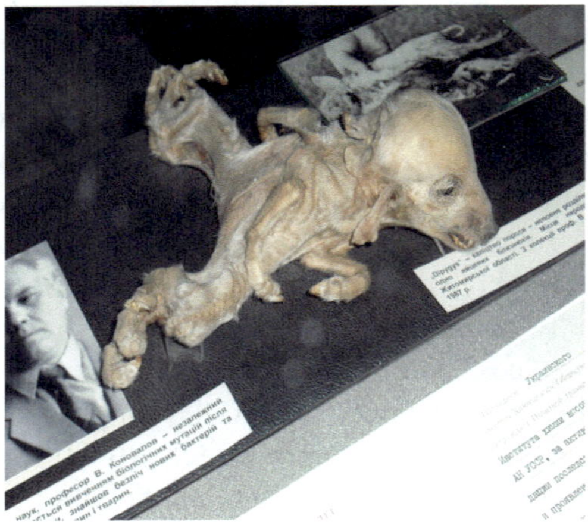

Figure 7.8. *Chernobyl nuclear disaster effects.*

Source: Vincent de Groot – http://www.videgro.net, https://upload.wikimedia.org/wikipedia/commons/0/02/Kiev-UkrainianNationalChernobylMuseum_15.jpg, Creative Commons License.

Distinct channels housed 222 control and shutdown rods, and the reactor was capable of being refueled online by a refueling apparatus positioned above the core. The operators faced pressure to conclude the test due to the subsequent maintenance time being over a year away, a scenario exacerbated by the request for an 8-hour delay. The emergency core cooling system was disabled for the test. The initial phase of the test involved decreasing the power from 3,200 MWt to around 800 MWt; however, it was lowered to 30 MWt, which was insufficient to counteract the accumulation of ^{135}Xe fission product poison. The swift accumulation of ^{135}Xe rendered it exceedingly challenging to restore the power level. The operators, in contravention of all regulations, removed the majority of the control rods in an effort to counteract the ^{135}Xe poison and rectify their previous error; however, they failed to restore the power level to 800 MWt and could not exceed 200 MWt. The RBMK reactor exhibits instability at 200 MWt, a condition exacerbated by the deactivation of multiple safety mechanisms to avert circuit interruptions during the experiment. Under these circumstances, steam voids commenced to develop in the coolant circulating within the pressure tubes, resulting in increased positive reactivity due to a diminished presence of hydrogen atoms to absorb neutrons.

Figure 7.9. *NRX reactor.*

Source: https://upload.wikimedia.org/wikipedia/commons/8/83/NRX_Pile_Building_and_ZEEP_Building-_Cooling_Tanks_1945.jpg, Creative Commons License.

The positive reactivity coefficient constituted a critical design defect. An elevation in temperature diminishes the density of the cooling water within the pressure tubes, leading to a reduced absorption of neutrons by hydrogen atoms. But, these additional neutrons would remain thermalized within the graphite moderator; hence, every increase in fuel temperature results in a net escalation of fission events, an effect that proliferates swiftly unless neutron absorption is promptly introduced. Even under standard conditions, a complex array of detectors, circuits, and control rods was necessary to regulate the power level in the RBMK reactor (Babenko et al., 2007). The formation of steam voids resulted in an abrupt increase in positive reactivity. This resulted in a power surge of around 30,000 MWt, 10 times over normal, and it could not be rapidly mitigated due to an excessive number of rods being withdrawn to exert any influence. The power surge destroyed the fuel, and the elevated steam pressure fractured the coolant tubes. Chemical reactions involving steam, graphite, and zirconium generated large quantities of carbon monoxide and hydrogen, resulting in an explosion that breached the roof, which was intended for confinement but not containment. The heated graphite ignited and burned for several days, causing the core to dissolve. A large quantity of fission products was elevated into the air, dispersing over Eastern Europe. Emergency workers were exposed to radiation levels ranging from 100 to 1,500 rems, resulting in the deaths of 31 individuals, mostly firefighters. Thousands of individuals were evacuated, including 45,000 from the town of Pripyat, many of whom were permanently displaced at substantial expense. The majority of these individuals received fewer than 25 rems. The specific amount of radiation exposure to workers and the public, estimated at 1.6 million person-rems, remains uncertain; however, the collective dose to the public is predicted to substantially elevate cancer risk. Many variables led to the Chernobyl disaster, but the critical defect was the positive void coefficient, which has since been modified in other reactors of the same design in the area.

7.13. NRX REACTOR

The NRX reactor (Figure 7.9) was developed in Canada during World War II to demonstrate the application of D_2O, or heavy water, as a moderator for a reactor utilizing natural uranium as fuel. It is distinguished as the precursor to the highly beneficial CANDU nuclear power reactor.

Figure 7.10. *K-reactor.*

Source: https://upload.wikimedia.org/wikipedia/commons/2/2a/KSTAR_tokamak.jpg, Creative Commons License.

The uranium rods in the NRX were encased in a pressure tube and cooled by a small layer of light water circulating between the aluminum-clad fuel rods and the wall of the pressure tube, which is a larger concentric aluminum cylinder. The heavy water moderator circulating between the pressure tubes adequately moderated all neutrons to thermal energy; hence, the light water coolant acted as a neutron absorber, reducing the quantity of neutrons available for fissioning uranium atoms. The NRX accident emerged during studies necessitating a diminished flow of light water coolant for many fuel rods. Operator errors, along with failures in electrical and mechanical safety circuits, also led to the tragedy. At an output of approximately 17 MW, the cooling water began to boil in the channels with diminished coolant flow, which effectively eliminated some of the water neutron-absorbing properties due to the reduced density, resulting in an increase in power once again. The power escalated to 60 MW and subsequently to 90 MW, at which juncture the heavy water moderator was automatically discharged, halting the reaction. Approximately 1.2×10^{20} fissions had developed, resulting in damage to the core and calandria. Approximately 104 Ci of long-lived fission products were discharged into the basement with roughly 10^6 gallons of cooling water that required clean up. Personnel exposures were minimal, and the reactor was recommenced approximately one year later. The NRX incident demonstrated the beneficial reactivity effect when light water is employed to cool pressure-tube fuel channels, a scenario that bore similarities to the RBMK Chernobyl reactor. This serves as another illustration of the damage produced by the steam explosion that occurs after a sudden power spike. It is significant that the fission reaction can be halted by displacing the D_2O moderator, a benefit that a heavy water reactor possesses that a graphite-moderated reactor lacks. A catastrophic nuclear criticality incident, resulting in three fatalities, transpired at the stationary low-power unit one (SL-1) reactor in Idaho on January 3, 1961. The reactor was engineered for the US Army to supply power in isolated areas, with minute containment systems for enhanced portability. The reactor was deactivated for repair, and three operators arrived for the night shift to reattach the control rods, enabling the reactor to be reactivated the following morning. It was essential to physically elevate the control rods to reestablish their connection. The precise circumstances remain unclear; but, it is presumed that the control rod was immobilized, and upon exerting force, it released a significant and abrupt surge of reactivity. The steam explosion ejected all control rods and elevated the massive reactor vessel approximately 9 feet.

7.14. K-REACTOR

K-reactor (Figure 7.10) is one of several heavy water-moderated reactors constructed in the early 1950s to generate tritium through neutron irradiation of lithium (^6Li) targets (Goncharov, 1964).

The reactor functioned effectively for approximately 25 years until being decommissioned for several safety enhancements, chiefly to augment seismic stability. The renovations required approximately one year, exceeding initial expectations, during which tritium in the targets decayed to ^{3}He, a prominent neutron absorber with a cross-section of 5330 b. Upon restarting the reactor, the operators had to retract the control rods far more than usual. Upon this observation, the Department of Energy officially mandated the reactor shutdown until this anomalous condition was fully assessed.

Figure 7.11. *Fukushima-Daichi facility.*

Source: https://upload.wikimedia.org/wikipedia/commons/0/07/Fukushima_Daiichi_04780015_%2883881740 45%29.jpg, Creative Commons License.

At this juncture, the deleterious impact of the ingrown ^{3}He was ascertained; however, more investigations regarding the procedural formalities for identifying and evaluating anomalous conditions, including self-imposed ones without prior evaluation of their effects on reactor criticality, were prompted. The K-reactor remained inactive for several years post-1988, was reactivated following expensive equipment improvements, and subsequently entered a standby status. The resumption of operations at this old reactor is uncertain; alternative methods for tritium production for nuclear weapons are expected to be developed.

7.15. FUKUSHIMA-DAICHI FACILITY

The Fukushima-Daichi facility (Figure 7.11) comprises six Mark I boiling water reactors produced by General Electric (GE) (Von Hirschhausen, 2018). The Mark I design is an initial iteration of a Boiling Water Reactor (BWR) including a water-filled torus that encloses and connects to the pressure vessel, serving as a pressure suppression pool in the event of accidents. Multiple reactors were operational on March 11, 2011, when a 9.0 magnitude earthquake struck many kilometers offshore. The earthquake was detected at the facility, prompting the automated shutdown of operational reactors, in accordance with standard protocol. However, the earthquake caused a massive tsunami around nine meters in height, which reached the location a few hours later, flooding seawalls, reactors, and draining all offsite power. Despite each unit being equipped with backup diesel generators to maintain system and cooling pump functionality, these generators were situated below ground level and were entirely inundated by the tsunami. Operators advanced the provision of offsite power for the pumps essential for dissipating residual heat from the core and the used fuel pools. The fuel overheated, resulting in a water-metal reaction that oxidized the fuel coating, releasing hydrogen that accumulated in the containment. Operators seemed hesitant to release the hydrogen because of the uncertainties of the surrounding population regarding radiation and radioactivity stemming from the bombings of Hiroshima and Nagasaki. As a result, a volatile concentration accumulated, leading to hydrogen explosions on March 14, approximately three days later, which detached the roof and walls of units 1 and 3, subsequently releasing airborne radioactive material. Radioactivity levels exceeding specified limits were observed in salmon, milk, and spinach several days later. An electric cable feed was established on March 19 to supply power to the cooling pumps at two reactors, approximately eight days following the automatic shutdowns. By that time, considerable damage to the fuel had emerged in both the core and the spent fuel pools, despite the efforts of plant operators, which included the utilization of seawater for emergency cooling. The seawater adversely affected all systems, particularly the reactor fuel within the core and the spent fuel pools, which housed between 50 and 151 tons of spent fuel, raising concerns about the potential loss of fuel cooling. Water levels in the pools were diminished to around 50% at reactors 1 and 3, while remaining elevated at reactor 2. Reactor fuel was introduced to the pool at reactor 4, resulting in increased thermal loading that vaporized water and subjected the zirconium-clad fuel to an exothermic reaction that might transpire to 1800 degrees Celsius. The pumps were restored, and the fuel pools were secured; after approximately one year, all radioactive fuel had cooled sufficiently to eliminate the likelihood of melting. A significant portion of the discharged material was dispersed offshore and away from densely inhabited regions. The Fukushima-Daichi tragedy reiterates the necessity for the nuclear sector to have reliable emergency power to ensure continuous fuel cooling. Preventing the criticality of fissile material relies on the same considerations that must be addressed to get a supercritical state in regulated nuclear reactors or nuclear weapons. Criticality safety is based on the following factors: (1). Restricting the amount of fissile material to below the critical mass in all possible configurations; (2). Limiting the geometry of the fissile material to one that is always safe, meaning a configuration where the surface-to-volume ratio ensures significant neutron leakage, rendering it difficult to achieve a multiplication factor of 1 or above; (3). Restricting the concentration of any fissile material inside a solution of moderating substances, such as H_2O, D_2O, or other liquids, while taking into account the presence or absence of neutron-reflecting materials; and (4). Avoiding the interaction among multiple subcritical arrays of fissile materials, particularly during transit and storage. The effects of mass and density impact the critical mass of fissile material. The critical mass of a sphere of ^{239}Pu metal is significantly lower than that of a sphere composed of unmoderated ^{239}Pu filings. Nuclear weapon designs exploit this principle by compressing a spherical mass to a high density.

7.16. GEOMETRIC EFFECTS

Geometry effects can be employed in a consistently secure configuration to avert unintentional criticality by facilitating neutron loss, even in the presence of adequate mass and/or moderation and reflection (Poston et al., 2020). When a specific volume of fissile solution deviates from a spherical configuration, the surface area increases, allowing more neutrons to escape. A crucial geometric consideration is the spacing of subcritical units to prevent mutual neutron exchange. This separation is particularly crucial in facility design as well as in the storage and transportation of subcritical units. The moderator and reflector effects of light materials, particularly water, can significantly affect the critical configuration of fissile material, decreasing its critical mass to merely a few 100 grams. Highly dilute solutions where neutron absorption by hydrogen is predominant can accommodate high amounts of fissionable material without achieving criticality. A densely arranged subcritical array may achieve criticality when flooded; a submerged subcritical array of larger, less densely packed units may attain criticality upon the removal of water, as the water acts as a neutron absorber, hence reducing neutron coupling among the units. Subcritical systems can transition to criticality if a reflector, such as beryllium, water, or any biological tissue, is introduced, therefore returning escaping neutrons to the array. Most scenarios are not accurately depicted by infinite or spherical slab geometries, nor are they entirely regulated. A larger quantity of fissionable material may be present without achieving criticality; however, permitting these conditions requires the highest degree of caution. The separation of subcritical units is a crucial factor in evaluating criticality safety, particularly if one unit could significantly contribute neutrons to the other. Such separation is crucial between 'always safe' containers and pipelines to prevent them from attaining a critical configuration, and caution must be exercised when positioning materials that are effective neutron reflectors. Densely constructed concrete serves as an effective reflector; hence, positioning vessels containing concentrated fissionable material adjacent to a wall or floor of such concrete may induce subcritical concentrations to reach criticality. Containers that provide consistent safety spacing are typically referred to as bird cages. A 55-gallon barrel with a unit container welded centrally using suitable spacers constitutes a birdcage. The drum serves as the spatial container, ensuring that the minimum distance between the surfaces of fissile materials remains below a safe separation threshold, even when the containers are submerged in water. Criticality safety protocols restrict at least one of the elements influencing criticality to values below the designated minimums. It is advisable to depend on equipment designs rather than primarily on processes to guarantee safety. A process intended to satisfy a concentration or enrichment condition may fail if inappropriate material erroneously enters the process, a situation that can be mitigated through equipment designs that ensure a consistently safe geometry. Restricting the enrichment level of fissionable materials in a liquid concentration is also an effective measure. An aqueous mixture or solution of uranium cannot achieve criticality unless the concentration of ^{235}U exceeds 0.95%; so, there are no criticality restrictions for the transportation or storage of natural uranium or homogeneous aqueous mixtures containing less than 0.95% ^{235}U. Uranium metal enriched to less than 5% ^{235}U cannot attain criticality, assuming it is not arranged in rods within hydrogenous material. Criticality events produce fission products alongside a significant exposure to neutrons and gamma rays. Volatile fission products, including Iodine, bromine, and noble gases, typically escape due to the generated heat, potentially resulting in airborne concentrations within structures or their release into the environment. Others persist in the medium where the criticality emerged and can be utilized to quantify the event, typically in relation to the number of fissions that occurred.

A calculation of fission product releases from criticality events must first acknowledge that these events emerge instantaneously, implying that any gaseous and volatile fission products available for release are those generated instantaneously, as determined by their independent fission yields. The inventory will inevitably evolve over time as the fission products undergo transformation and progress along the fission product chain, occasionally resulting in the creation of additional gaseous and volatile radionuclides. The volatiles generated directly will not produce their outcomes from the radioactive decay linked to fission product chains. The absorption of radiation in a gas-filled chamber is achieved by creating an electric field between the chamber wall and a positively charged electrode positioned along the chamber axis and insulated from it. The gas typically exists at a pressure of one atmosphere or lower, but can be compressed with proper precautions to improve interactions. When radiation is absorbed by the gas within the chamber, ion pairs are generated, which are subsequently collected and amplified for signal recording. The accumulation of ion pairs in a gas-filled detector is dependent on the applied voltage, which delineates the performance of the detectors for beta particles and alpha particles. In the absence of voltage across the chamber, the ion pairs will recombine, resulting in no charge flow in the external circuit for either radiation type. As the voltage is elevated, for instance to several volts, certain ion pairs will continue to recombine while others will migrate to the electrodes. At a voltage of around 10 V or higher, recombination becomes insignificant, allowing all electrons generated by ionization to reach the core electrode. When the voltage is elevated to several volts, the quantity of ion pairs collected or deposited remains constant regardless of the applied voltage, and the curve will remain horizontal as long as the radiation source emits ionizing radiation at a consistent pace. When 10 ion pairs are initially generated, the response remains stable, as indicated by the lower curve. If the ionizing rate increases by a factor of 10, resulting in the formation of 100 ion pairs, the detector response will also increase by a factor of 10.

Figure 7.12. *Geiger-Mueller (GM) counter.*

Source: https://upload.wikimedia.org/wikipedia/commons/4/40/Geiger_counter.jpg, Creative Commons License.

The curves are parallel, indicating that the collected current is directly proportional to the ionization generated by the incoming radiation. The area where the measured current correlates directly with the quantity of ionization generated is termed the ionization region; its intensity will vary significantly for equivalent source strengths of alpha particles, beta particles, X-rays, and gamma rays due to the differing numbers of ion pairs produced by each during their interactions within the medium. Elevating the voltage of a gas-filled detector beyond the ionization area enables the electrons generated by primary ionizations to gain sufficient energy to induce further ionizations upon colliding with gas molecules within the chamber. The quantity of captured electrons increases almost exponentially with the applied voltage, as each starting electron is accelerated, generating a minor avalanche of electrons, predominantly released around the core electrode. Each electron released by the incoming radiation generates an individual avalanche, resulting in a constant amplification of ionization at a specified detector voltage; so, the quantity of collected ion pairs is proportional to the ionization. Gas-filled detectors functioning at these voltages are classified as operating in the proportional region. As the voltage across the chamber attains several 100 volts, the gas multiplication effect escalates rapidly, leading to increased electron avalanches that subsequently interact, resulting in a region of limited proportionality. As the voltage is further elevated, the accumulated charge becomes independent of the initial ionization, resulting in the two curves not only becoming identical but also forming a plateau with increasing voltage.

7.17. GEIGER-MUELLER (GM) COUNTER

The GM counter (Figure 7.12) functions within the Geiger area or region and is distinguished by a plateau voltage that generates an avalanche of discharge for each instance of ionizing radiation entering the chamber. The surge of charge generates a pulse, the magnitude of which is unaffected by the initial ionization; hence, the GM counter is particularly effective for detecting low-ionizing radiations such as beta particles or gamma rays and is specifically engineered to exploit this phenomenon.

Figure 7.13. *Galvanometer.*

Source: https://wellcomeimages.org/indexplus/obf_images/30/af/06570418c9a34ae5c68c689cd90a.jpg, Creative Commons License.

Due to the challenge of fabricating tubes with sufficiently thin windows for alpha particles to enter the gas chamber, GM counters are utilized for detecting the penetrating beta and gamma radiations. A thin window is frequently integrated into GM tubes to improve the detection of beta particles. A GM tube typically has a thin wire electrode, such as tungsten, with a mixture of 90% argon and 10% ethyl alcohol at a sub-atmospheric pressure. A potential difference ranging from 800 to 2,000 V is applied to render the tube negative in relation to the wire. The gas-filled radiation detectors react to incoming radiation by the ionization it generates. The ions, whether increased in quantity or not, are collected to generate a voltage pulse that can be as little as 10 μV. The tiny pulses are amplified to 5 to 10 V and transmitted to a galvanometer (Figure 7.13) for direct measurement or to a pulse counter for analysis.

The intensity of the generated pulses can be quantified in the proportional region and categorized based on their energy; alpha particles vs beta particles. Diverse crystals and solid-state detectors can be employed with electronic instruments to quantify the number of radiations of a specific energy that activate the detector and the frequency of such incidences. A crystal like sodium iodide with a thallium dopant, NaI(Tl), generates light that is proportionate to the energy imparted by a photon interacting within the crystal (Maddalena et al., 2019). Light, defined as a photon with a wavelength within the visible range of electromagnetic radiation, is reflected onto a photocathode linked to a photomultiplier (PM) tube. Upon striking the light-sensitive cathode, these photons cause the emission of a limited quantity of electrons, proportionate to the energy of the absorbed photon. A potential difference across the first dynode accelerates these, resulting in the emission of around four secondary electrons for each incident electron. A sequence of 10 dynodes yields an amplification factor of 4^{10}, or around one million. Each pulse emitted from the 10th dynode correlates with the energy absorbed, which is directly proportional to the energy of the photon impacting the crystal. The pulses are categorized by size and recorded in the respective channels of a multichannel analyzer as counts, enabling the identification of the source based on its distinct gamma energy. Sodium iodide crystals have notable efficiency in gamma ray detection due to their density and the large atomic number of iodine, with a pulse resolving time of approximately 0.25 microseconds, enabling their application in high count rate scenarios. Plastic scintillators possess shorter resolving periods; but, their photon detection efficiency is suboptimal, and the light output relative to energy is non-linear, leading to spectra with diminished resolution. The energy deposition in a semiconducting material like pure germanium excites electrons from filled valence bands to conduction bands, resulting in the formation of conduction electron pairs and electron vacancies, or holes. A bias voltage is provided to the semiconductor, inducing the movement of charge carriers and generating a current pulse. The energy required to generate an electron-hole pair in a semiconductor is often approximately 1 eV, significantly lower than that necessary for ionizations in a scintillator; hence, a reasonably large quantity of charge carriers is generated for each absorbed photon. The statistical variations in the number of atoms excited or ionized are significantly reduced in a semiconductor detector. When employed with a multichannel analyzer, this yields sharp peaks or high resolution, enabling precise energy determination. This technology has progressed to a point where high-efficiency germanium detectors are consistently manufactured, offering exceptional photon resolution.

7.18. SEMICONDUCTOR DETECTORS

Semiconductor detectors possess many advantages compared to scintillators. They exhibit high efficiency in photon detection due to their density and generate rapid pulses with exceptional resolution.

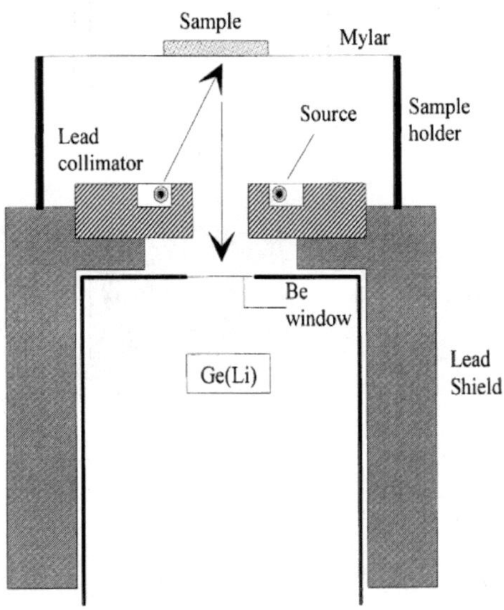

Figure 7.14. *Ge(Li) detectors.*

Source: Durak, R., & Özdemir, Y. (2001). X-ray fluorescence escape peaks in Ge(Li) detectors. Instrumentation Science & Technology, 29(3), 185–192. https://doi.org/10.1081/CI-100103465, Creative Commons License.

Germanium and lithium-drifted germanium (Ge(Li)) detectors (Figure 7.14) necessitate operation at liquid nitrogen temperatures to mitigate thermal noise, presenting a potential drawback (Darken and Cox, 1995). Ge(Li) detectors rapidly degrade if not maintained at liquid nitrogen temperatures due to the diffusion of lithium into the germanium matrix. Ge(Li) detectors have been replaced by intrinsically pure germanium detectors, which can be stored at ambient temperatures; however, they require cooling during operation. Alternative semiconductor detectors consist of silicon infused with lithium to detect electrons and alpha particles, and when paired with pulse height analyzers, they facilitate spectral analysis. These detectors offer greater resolution and can be maintained at ambient temperatures due to the reduced mobility of lithium in silicon compared to germanium.

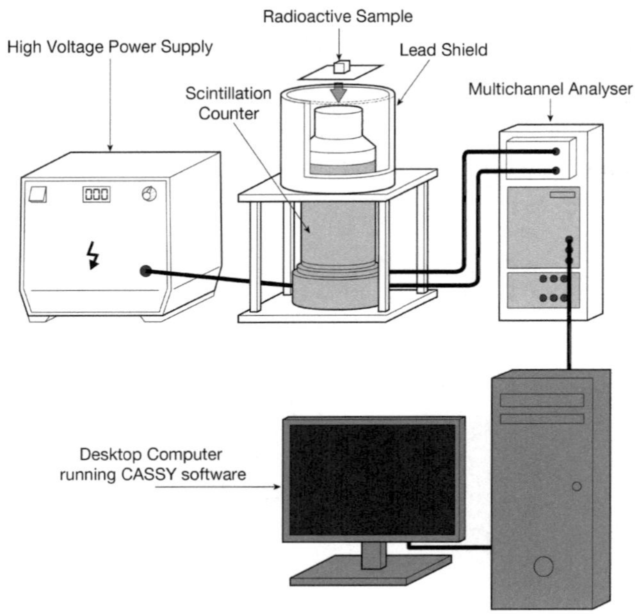

Figure 7.15. *Gamma ray spectrometers.*

Source: https://upload.wikimedia.org/wikipedia/commons/a/ac/Scintillation_counter_as_a_spectrometer.jpg, Creative Commons License.

Gamma ray spectrometers (Figure 7.15) employ photon detectors engineered to exploit diverse absorptive processes inside a high-Z medium to enhance sensitivity and deliver spectral data regarding the collected photons. The intricacies of photon interactions are apparent in the energy spectra of gamma emitters. The distinctive characteristics can be broadly categorized into those that manifest below the threshold for pair production and those that rise above this threshold. A NaI(Tl) crystal is a widely used scintillation detector that generates light scintillations proportional to the energy of absorbed photons. The scintillations of light result exclusively from ionizations and excitations caused by electrons released during photon interactions within the crystal. Due to their charged nature, nearly all emitted electrons are absorbed within the crystal, generating an output pulse that reflects the quantity of absorbed photon energy, thereby quantifying the interacting photons and their energy levels.

Complex spectra arise when the energy of the initial photon is adequate for absorption via pair production, specifically when hm $\geq$ 1.022 MeV. In the case of a large detector, the full-energy photopeak will be observed; however, additional characteristics will also be evident, recorded using an inherently pure germanium crystal. Sodium-24 generates two high-energy gamma rays at 1.369 MeV and 2.754 MeV, resulting in photopeaks at these energies in the gamma spectrum. All principal photon interactions can emerge for Na-24 gamma rays; however, the predominant interactions are Compton scattering and pair production, as the photoelectric absorption coefficient in germanium is minimal at this elevated energy. When Compton interactions occur and all subsequent processes are finalized during the span of the crystal, two photopeaks will be produced at 1.369 and 2.754 MeV. The peaks result from pulses that correspond to the complete deposition of photon energy through many processes occurring within the crystal. A Compton continuum of pulses arises from the absorption of Compton electrons beneath each photopeak, as certain Compton-scattered photons are lost from the crystal, with only the energy transmitted to each ejected electron being absorbed and recorded. Gamma emitters that facilitate pair production provide a prominent peak at 0.511 MeV.

CHAPTER 8

RADIATION PHYSICS AND METALS

Contents

8.1. INTRODUCTION

When a metal atom absorbs energy from a moving particle that meets or exceeds a certain threshold value, it is displaced from its location inside the crystal lattice. The energy of the knocked-out atom, influenced primarily by the type and energy of radiation, the characteristics of the irradiated target, the crystallographic orientation, and the angle of particle collision with the lattice atom, can result in diverse configurations and structures of radiation defects.

These range from isolated point defects such as vacancies and interstitials or Frenkel pairs to extensive damage zones exhibiting complex defect structures.

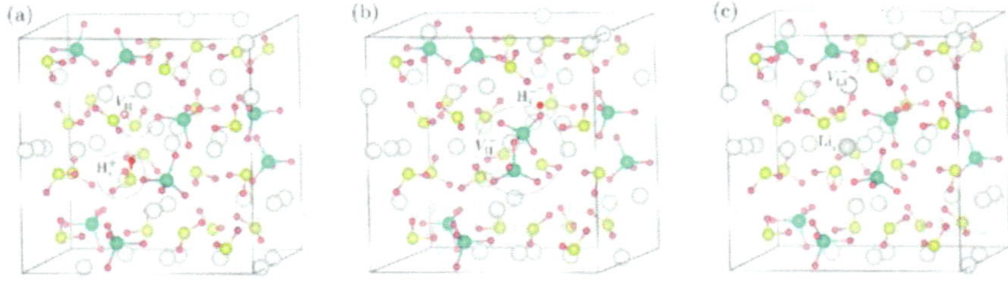

Figure 8.1. *Frenkel pairs (a,b,c).*

Source: Hoang, K., Janotti, A., & Walle, C. G. V. de. (2014). The role of native defects in the transport of charge and mass and the decomposition of Li4BN3H10. Physical Chemistry Chemical Physics, 16(46), 25314–25320. https://doi.org/10.1039/C4CP03677H, Creative Commons License.

Single Frenkel pairs (Figure 8.1) are generated with energies of the knocked-out atoms within the range which is for the irradiation of copper with electrons possessing energies below 1 MeV (Kozlovski et al., 2015). At excited or elevated energies of the ejected atoms, a cascade of atomic collisions occurs, resulting in the development of a complex configuration of radiation defects. The interactions between radiation and the crystal lattice of metals in the creation of radiation defects, as well as the methodologies for modeling these interactions, have been thoroughly analyzed. The faults arising from the interactions between moving particles and matter constitute primary radiation defects. The ensuing kinetic progression or evolution of radiation damage structure is governed by the spatial overlap of initial damage volumes and/or the diffusion of freely migrating point defects and solutes. A series of atomic collisions occurs at elevated energy. At elevated recoil energies, vital atomic collisions, as well as the production and recombination of Frenkel pairs, occur near the trajectory of the primary-displaced atom, which constitutes the center of the cascade. This process involves the creation of chains of concentrated atomic collisions, resulting in the segregation of vacancies and interstitials within the Frenkel pairs, and the localization of interstitials at the cascade periphery. Chains of atomic displacements were originally identified using computer modeling of atomic collisions in copper, utilizing the molecular dynamics approach. The ultimate configuration of the cascade comprises a neutral zone with an elevated concentration of vacancy-type defects, and a periphery distinguished by the presence of isolated interstitials or clusters of these atoms. Intra-cascade recombination, which defines cascade efficiency for analogous models, can be assessed using the analytical equations. At particular energies of the primary-displaced atom, cascade branching may occur. The increase in recoil energy to several KeV in studies involving several particle types demonstrates a significant reduction in defect creation efficiency, followed by a tendency towards saturation. The discrepancy between the theoretical and experimental results is clearly linked to the mutual recombination of point defects, a phenomenon that is challenging to incorporate into analogous models, particularly at the formation energies of subcascades. The primary distinctive characteristics of cascade growth and the configuration of the damaged zone, analyzed within the context of the simple binary collision model, are in qualitative agreement with the outcomes of computer simulations performed using the molecular dynamics approach. Simultaneously, recent research utilizing computer modeling at elevated energies of primary recoil atoms reveal a significant characteristic of cascade growth, called the emergence of a thermal peak. Within the thermal peak range, temperatures can significantly exceed the melting point. The realistic creation of the molten zone in computer modeling experiments has been validated by assessments of atomic kinetic energy, atomic density in the thermal peak region. The rapid rise in temperature within the thermal peak zone significantly amplifies the mutual recombination of defects. The emergence of a molten zone results in the creation of interstitial clusters at the cascade periphery, attributable to two mechanisms: the ballistic displacement of interstitials from the molten zone and the cluster formation induced by a shockwave generated during the rapid cooling of the molten zone. The escalation of mutual recombination of point defects during the emergence of a thermal peak is a significant outcome of computer modeling, since it aligns with practical observations.

8.2. SUBCASCADES

The development of subcascades, first detected in computer modeling tests with a recoil energy of 25 keV in, was accompanied by the merging of the molten zones of two subcascades (Stoller and Greenwood, 1999). There has been a tendency to compare the efficiency of structural phase changes induced by various types and energies of radiation, focusing on the assessment of the concentration of freely migrating defects that do not recombine or merge into clusters during primary radiation damage. This criterion clearly possesses a distinct advantage over the dpa parameter, particularly at high temperatures, which are characterized by swift diffusion-controlled processes including the nucleation and development of dislocation loops, clusters, voids, and various phase transitions. The proportion of freely migrating point defects can be assessed by the theory of binary collisions in the modified Kinchin–Pease model or via direct computer simulations of atomic collision processes. The extent of spatial correlation among defects escalates in this instance. It is clear that for neutrons, despite their minimal mass, the spectrum of primary-displaced atoms is comparatively hard. This is linked to the phenomenon that atomic displacements during neutron radiation occur at minimal impact parameters. When comparing ions that can transfer both substantial energy during direct elastic collisions and minimal energy through Coulomb interactions over distances, the proportion of defects generated in energetically dense cascades is greater for neutrons. The narrow energy spectrum of recoil is produced by 1 MeV electrons, which are characterized by the production of isolated Frenkel pairs. These considerations are relevant in the qualitative analysis of the experimental data acquired during the assessment of the fraction of stable defects. The fraction of freely-migrating defects can be determined by the examination of results from both direct and indirect tests. The indirect tests encompass investigations conducted at temperatures where point defects remain stationary or stable.

8.3. TEMPERATURE DEPENDENCY

At higher radiation temperatures, the concentration of freely moving defects, at a recoil energy above 25 keV, characteristic of fission and fusion neutrons, diminishes to roughly 1% (Stoller and Greenwood, 1999). A qualitative analysis of the temperature-energy relationship was conducted, suggesting the enhancement of intra-cascade recombination and the processes of recombination and defect cluster creation in the vicinity of adjacent cascades. Researchers believe that interpreting the temperature-energy dependency necessitates considering the impact of thermal oscillations of the lattice atoms on defect formation kinetics. Modeling atomic collisions in Cu reveals that the average distance between the vacancy and the interstitials in Frenkel pairs is nearly reduced by half as the irradiation temperature rises from 0 to 293 K. This correlates with a likelihood of defocusing in sequences of substitutive atomic collisions alongside an increase in the amplitude of atomic oscillations. Studies demonstrate a satisfactory correlation between the experimental and calculated data on the assessment of the fraction of freely-migrating defects. The models evaluate the fraction of freely-migrating defects, focusing on the development of stable isolated Frenkel pairs originating from both the primary-displaced recoil atom and subsequent secondary atomic collision generations. It was suggested that stable isolated point defects arise from sequences of atomic collisions whose separation surpasses the contact radius of point defects for their recombination or complex development. Particular emphasis was placed on the influence of impurities and alloying elements in pure metals on the development of radiation defects, including defect-impurity complexes, defect clusters, and voids. The issues related to phase transitions and their impact on the pattern of radiation damage have been examined superficially. The creation processes of defect clusters, dislocation loops, and voids are typically interconnected with fluctuations in the chemical composition of the matrix solid solutions and may also link to the phase instability of both single-phase and more complicated alloys. The comprehensive analysis of these mechanisms facilitates an efficient interpretation and prediction of radiation damage in alloys. The current interest in radiation-induced phase transitions is linked to enhance the performance parameters of materials in nuclear power engineering that include temperature, intensity, fluence, and durability of steels and alloys, as well as developing fusion reactors where the operational conditions for structural metallic materials under irradiation, coupled with other factors, approach extreme conditions. Radiation-enhanced diffusion plays a crucial role in the amplification of structure-phase processes in solid solutions and increasingly complex metallic systems. Volume radiation-enhanced diffusion is a primary mechanism of radiation-induced phase transitions, advancing the system evolution toward equilibrium. This pertains mainly to the expedited merger of phases due to irradiation, governed by the diffusion of solutes inside matrix solid solutions.

8.4. MATERIALS

The resistive materials, particularly solid solutions, may experience alterations in their physical properties, particularly electrical resistivity, due to radiation-enhanced interdiffusion. It is evident that without irradiation at temperatures below 80–70°C, the equilibrium values of electrical resistivity under actual thermal conditions are unfeasible; but, with irradiation, they can be achieved. A similar temperature dependence of electrical resistivity change was demonstrated for the Cu + Zn alloys. It is important to note that the acceleration of various processes toward thermodynamic equilibrium during irradiation, including ordering, may be supplementary and linked to the compensation of coherent strains generated during transformations by point defects. Preliminary deformation mitigates the impact of radiation during diffusion processes, analogous to the phenomenon observed in swelling. For steels utilized in the active zones of nuclear reactors, the technique called as plastic deformation is employed to diminish the extent of swelling. Under identical irradiation and migration energy parameters for point defects, the interdiffusion coefficient attains its maximum value. The coefficients of radiation-enhanced diffusion of solutes in diluted alloys via the interstitial mechanism can be ascertained using two models: the Barbu model and the Bocquet model. The Bocquet model is also beneficial for the approximate assessment of the coefficients of radiation-enhanced diffusion via the interstitial mechanism in concentrated alloys

(Allen and Was, 2007). The majority of experimental research on radiation-enhanced diffusion were conducted through indirect investigative approaches to analyze short-range ordering and clustering processes in alloys subjected to irradiation. The research was conducted on the silver-zinc, copper-zinc, copper-aluminum, copper-nickel, and gold-silver systems utilizing electrical resistivity measurement and Zener relaxation methods.

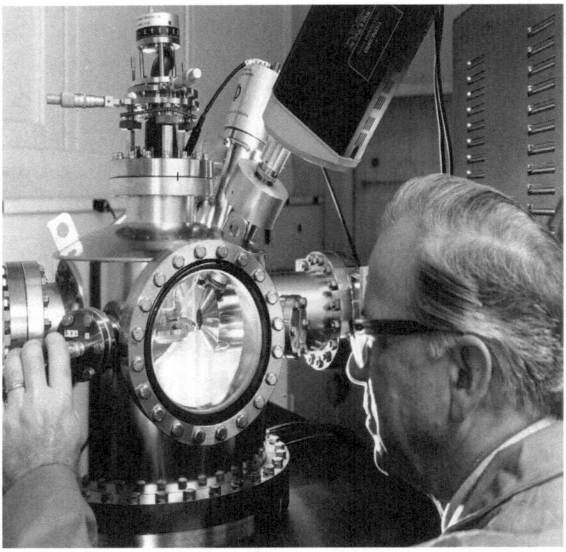

Figure 8.2. *Auger spectroscopy.*

Source: https://upload.wikimedia.org/wikipedia/commons/3/3f/HD.6C.037_%28 11856519893%29.jpg, Creative Commons License.

Fewer experiments of radiation-enhanced diffusion were conducted utilizing direct methods, including radioactive isotopes, Auger spectroscopy, (Figure 8.2) and mass spectroscopy (Figure 8.3). The outcomes of these trials on self-diffusion in silver, nickel in copper, copper in nickel, lead diffusion in silver, interdiffusion in nickel, and aluminum in nickel have been documented. The primary issues in analyzing the experimental results are to the accurate interpretation of the mechanisms underlying radiation-enhanced diffusion. The investigation of the mechanisms underlying radiation-enhanced diffusion within the system is conducted based on the temperature dependencies of the relaxation time of the measured characteristics. Considering that these alloys typically consist of concentrated solid solutions where the diffusion mobility of interstitials and vacancies may closely resemble that of pure metals, interpreting the mechanisms of radiation-enhanced diffusion solely based on the temperature dependencies of relaxation times or diffusion coefficients could yield misleading conclusions. This issue is equally significant for interpreting experimental data acquired through direct experimental procedures.

Figure 8.3. *Mass spectroscopy.*

Source: https://upload.wikimedia.org/wikipedia/commons/8/8d/Fenn_ESI_Instrument.jpg, Creative Commons License.

8.5. KINETICS

Based on the investigation of process kinetics, it is essential to assess the time required to achieve dynamic equilibrium, which is significantly smaller for the mutual recombination mechanism compared to the combined mechanism. Investigations indicate that the stationary kinetics of electrical resistivity fluctuation in this alloy at temperatures of –20 and –10°C is influenced by the minimal temporal variation of the coefficient of radiation-enhanced diffusion, attributable to the limited mobility of point defects. Experiments conducted using an electron microscope demonstrate that irradiation of the alloys within the temperature range of 80–250°C results in the formation of interstitial-type dislocation loops, while temperatures below 80°C induce the formation and growth of tetrahedral stacking faults of the vacancy type. The findings from quenching and electron microscopy experiments enable the interpretation of the activation energies for radiation-enhanced diffusion in this alloy, measured at 0.3 eV and 0.23 eV for low- and high-temperatures, as half of the activation energy for the migration of monovacancies and interstitials, respectively. The results indicate that, at temperatures below 70–80°C, the mechanism of radiation-enhanced diffusion is governed by the diffusivity of vacancies, while at elevated temperatures, it is governed by the diffusivity of interstitials. The transition temperature from the interstitial to the vacancy mechanism of radiation-enhanced diffusion is distinctly reflected in the temperature dependence of the equilibrium electrical resistivity values in this alloy. The kinetic analysis of the electrical resistivity variation in Ag-Zn alloys subjected to irradiation, which results in an increased density of dislocations at temperatures ranging from 10 to 35°C, indicates that the diffusion acceleration during irradiation is governed by a nonstationary combined mechanism. The kinetics of electrical resistivity variation is unaffected by the order of the reaction. The diffusion mobility of the vacancies is a determining factor. The shift from the interstitial mechanism of radiation-enhanced diffusion to the vacancy mechanism aligns with the relevant theoretical analyses. The effective mobility of interstitials in binary alloys is the aggregate of the mobilities of the dumbbell configurations of interstitials of various types, with their concentrations dependent on composition and temperature. The comparison of modeling concepts with empirical data from high-voltage electron microscopy allows for the assumption that at temperatures exceeding approximately 70–80°C, the processes of structural-phase transformations and the mechanism of radiation-enhanced diffusion in Ag – Zn alloys are governed by the migration of mobile interstitial dumbbell configurations of the Ag – Ag type, possessing an activation energy of 0.46 eV. A reduction in temperature is associated with the preferential production of less mobile dumbbell structures of the Ag – Zn and Zn – Zn types. The mobility of interstitials in the alloy significantly diminishes, and at certain temperatures, it falls below that of vacancies. The tests demonstrate this by the preferential creation and expansion of vacancy clusters, specifically tetrahedral stacking defects, at temperatures below approximately 80°C. Theoretical calculations and results from direct and indirect measurements indicate that, based on the radiation intensity and various other factors, radiation-enhanced diffusion in metals and alloys may dominate at temperatures up to 0.5–0.6 T_m. The coefficients of radiation-enhanced diffusion may vary from 10^{-20} to 10^{-14} $cm^2\ s^{-1}$. The radiation-enhanced diffusion in metallic materials can amplify structural-phase transformations at temperatures of 0.5–0.6 Tm, with radiation-enhanced diffusion coefficients ranging from 10^{-15} to 10^{-14} $cm^2\ s^{-1}$, depending on the irradiation intensity. The self-diffusion coefficients, solute diffusion coefficients, interdiffusion coefficients, and partial diffusion coefficients via the vacancy mechanism increase with irradiation in direct proportion to the degree of vacancy supersaturation. The enhancement of solute diffusion via the interstitial dumbbell mechanism is particularly effective for components exhibiting a negative dilation volume. A reduction in the coefficient of radiation-enhanced diffusion can be attained by increasing the concentration of diffusion sinks and by alloying with elements that diminish the disparity between the diffusion coefficients of interstitials and vacancies. The coefficients of radiation-enhanced diffusion can be approximately estimated based on models. The vacancy mechanism can also utilize experimental data derived from studies on phase transformations, thermal diffusion, and measurements of vacancy parameters in both equilibrium and nonequilibrium experiments. When bulk radiation-enhanced diffusion predominates, irradiated metallic systems approach thermodynamic equilibrium. The relaxation mechanisms linked to the elevated concentration of point defects, as well as dynamic processes such as chains of substituting collisions and displacement cascades, may significantly alter the equilibrium state of the irradiated alloys relative to their thermal equilibrium state. As irradiation temperature diminishes, the significance of radiation-enhanced diffusion diminishes, while the influence of dynamic processes resulting in disordering intensifies (Sizmann, 1978). The substantial influence is produced by the dynamic or ballistic impacts on structures exhibiting long-range order. A comprehensive analysis of the experimental data concerning the thermal and radiation ordering of the disordered Ni_3Mo system, utilizing computer modeling of the ordering processes within the nucleation domain and spinodal model, has enabled the identification of several distinctive characteristics of the long-range order variation under irradiation. A primary inference is that the irradiated system may exist in several states of equilibrium, depending on the ratio of the rates of ordering and disordering. The sequential progression of phases influenced by radiation conditions can be ascertained through an escalation in the amplitude of concentration of spinodal waves, the interaction of these waves with the development of domain nuclei, and the disintegration of domains when disordering.

8.6. THERMODYNAMIC STABILITY

The findings from experimental and theoretical studies indicate that dynamic or ballistic effects diminish the relative thermodynamic stability of structures exhibiting long-range order during irradiation below the critical temperature Tc, leading to the stabilization of the disordered state or the state with short-range order. The issues of disordering during irradiation are important due to the stability challenges of fine-dispersion precipitates in hardening ordered phases within industrial steels and alloys. Particular emphasis has been placed on the instability mechanism, based on the observation that the free energy of the disordered state exceeds that of the ordered particle. The disordered phase under irradiation exhibits a tendency for dissolution or transition to an alternative stoichiometric composition. Ion irradiation of Ni_3Si films at around

200°C resulted in the disorganization of the first phase and its conversion to the Ni_5Si_2 phase (Camacho-Olguín et al., 2020). The majority of experimental studies examining the impact of radiation on short-range clustering, utilizing measurement techniques such as transmission electron microscopy, electrical resistivity, diffusion, small-angle neutron scattering, and electron diffraction, have been conducted on the copper-nickel system. The investigations revealed that irradiation significantly accelerates both the short-range clustering process and the breakdown of the solid solution in the system. Analysis of various experimental data indicates that the copper-nickel system experiences short-range clustering and decomposition via the spinodal process during irradiation, leading to the creation of a modulated structure. Spinodal decomposition or breakdown resulting in spatially-oriented or spatially-modulated structures was also observed in matrix solid solutions of various steels and alloys within the Fe-Cr-Mn, Fe-Ni, Fe-Cr-Ni, and V-Ti-Nb-N systems. The interpretation of these results might also be grounded in previously investigated theoretical issues and several other theoretical frameworks. The analysis of radiation-induced breakdown in the iron-nickel system explains that the miscibility gap in this alloy significantly widens and shifts to higher temperatures during irradiation. In a series of investigations of the Cu-Ni-Fe system, characterized by breakdown via the spinodal mechanism, it was demonstrated that, depending on experimental conditions and parameters like temperature, particle type, energy, and intensity, irradiation may either facilitate the breakdown process or induce the dissolution of the precipitate due to cascade mixing. This transition is indicative of high-intensity ion irradiation and is unlikely to significantly impact structural reactor materials. In the early phases of radiation solid-state physics and materials science, research on phase transformations due to irradiation primarily focused on the kinetics and mechanisms of the decomposition of supersaturated solid solutions of copper, aluminum, and nickel, wherein the decomposition occurs via nucleus formation. Recent years have witnessed advances on radiation-induced phase transitions in similar systems. Fundamental investigations focus on the distinctive characteristics of radiation-induced breakdown and the relationships that dictate the change of Guinier–Preston zones into stable precipitates, depending on the settings and parameters of irradiation. In fundamental and applied research, the mechanisms of breakdown are examined in relation to the issues of radiation damage to materials of the specified grade. The primary theoretical models concerning the radiation-induced breakdown of supersaturated solid solutions, within the context of the mechanism for the production and growth of precipitate nuclei, have been established. Generally, they are appropriate for analyzing the phase instability of metallic systems subjected to irradiation, including under-saturated solid solutions. The theoretical analysis of nucleation and growth processes of non-coherent particles under irradiation has been conducted, considering the supersaturation factors of solid solutions by vacancies and solutes, as well as the dimensional mismatch parameters between the matrix solution and the precipitate. The evolution of precipitation is characterized by its trajectory in phase space, defined by the coordinates of vacancy concentration in a cluster and solute atom concentration. The process rate is dictated by the capture and emission rates of vacancies and solutes by the precipitates, as well as the absorption rate of interstitials.

8.7. VACANCIES

The method of elastic strain compensation at the matrix-particle interface and the presence of vacancies significantly influences the acceleration of precipitate nucleation and growth within the specified model. As vacancy supersaturation rises, the rate of solid solution disintegration escalates significantly. The model analyzes the clusters of solute atoms, which contain either an excess of interstitials or vacancies. The influx of solutes via vacancy and interstitial pathways to defect clusters with opposing charges, followed by the mutual recombination of defects, leads to a rise in the number of solute atoms within the clusters. This instance also considers the escape of solute atoms by interstitial and vacancy mechanisms. The model was employed to formulate an equation for the alteration of the solubility limit under irradiation conditions, depending on the ratio of the diffusion coefficients of the dissolved element via interstitial and vacancy mechanisms, as well as the parameter that dictates the relative likelihood of vacancy and interstitial atom localization within the clusters. The intensification of ordering, short-range clustering, and breakdown, which leads irradiated metallic systems toward a more equilibrium state, can serve as a preliminary irradiation treatment method for alloys, thereby stabilizing material properties. Researchers applied this technology to stabilize the electrical resistivity of resistive alloys, including Nichrome (Fe-Ni-Cu), Constantan (Cu-Ni), Ag-Pd, and Pd-Ag. The initial two systems pertain to short-range clustering, while the latter two are the ordering of metallic components. Initial electron or γ-irradiation of the resistors composed of these alloys led to the stabilization of their properties throughout operation in electronic systems, including space stations.

8.8. IRRADIATION

In the metallic materials subjected to irradiation, the acceleration of phase changes, a factor contributing to the destabilization of their properties, may also result in a reduction of radiation damage in the irradiated alloys. This is linked to the annihilation of point effects during the compensation of elastic strains, which occur alongside the processes of ordering, short-range clustering and breakdown, as well as the mutual recombination of defects at interphase boundaries. These methods reduce the dynamic equilibrium concentration of point defects, hence inhibiting the processes of nucleation and growth of defect clusters, dislocation loops, and voids. The radiation-induced breakdown of vanadium-titanium β-solid solutions, resulting in the creation of the α-Ti phase, is considered an important factor contributing to the high swelling resistance of vanadium-titanium system alloys. The results indicate that for particular wavelengths of the modulated structure, the minimal swelling during radiation-stimulated breakdown in a nickel-titanium alloy can be achieved. This phenomenon arises from radiation-enhanced diffusion, which is linked to an elevated concentration of point defects. Dynamic processes may significantly alter and even qualitatively transform the thermodynamic equilibrium conditions of irradiated alloys (Bellon, 2009). Certain mechanisms, including ordering, cascade mixing, and the compensation of elastic strains by point defects during phase transformations, have been addressed. We need to understand the processes governing phase instability and the displacement of phase equilibrium in metallic systems subjected to irradiation. Phase instability generally refers to the divergence of the phase composition of alloys from their actual or thermodynamically feasible equilibrium state in the absence of irradiation, as well as the alteration of the chemical composition of solid solutions under irradiation. Theoretical assumptions concerning the processes of phase instability during irradiation are now being developed based on thermodynamic and kinetic techniques. The assessment of phase equilibrium shifts under irradiation, grounded in the thermodynamics of equilibrium processes, primarily relies on the evaluation of the difference in free energies between the analyzed phases, attributed to variations in the concentration of point defects and the energy of interaction with the phase components. Similar estimations indicate that a significant displacement of phase equilibrium is feasible only if the concentration difference of point defects in the analyzed phases exceeds 10^{-4} to 10^{-3}. The cascade mechanisms, thermal peaks, and other processes may further alter the free energy of the irradiated material and facilitate a shift in phase equilibrium. The primary achievements in understanding phase instability under irradiation have been realized through the kinetic approach. The radiation-stimulated segregation processes involving impurity or transmutation elements may alter the chemical composition of primary precipitates. The initial segregation phase Mg_5Al_8 at voids partially transforms to the Mg_2Si phase due to the segregation of transmuted silicon atoms. In irradiated vanadium–titanium alloys, the synthesis of titanium compounds with carbon, oxygen, and nitrogen is preceded by the radiation-induced decomposition of the vanadium–titanium β-solid solution, resulting in the precipitation of the α-Ti phase. The subsequent separation of impurities from the α-Ti phase precipitates results in its conversion into titanium oxides or carbonitrides, characteristic of irradiated vanadium-titanium alloys. The reflections from the α-Ti phase precipitates are observed at relatively low radiation doses and are undetectable with subsequent irradiation. The mechanism of radiation-stimulated segregation, similar to other mechanisms of phase instability, can substantially influence the radiation resistance of structural metallic materials by altering the efficacy of sinks and the chemical composition of matrix solid solutions. This induces variations in the dynamic concentration of point defects and simultaneously impacts the processes of nucleation and growth of clusters, dislocation loops, and voids. The segregation of various types of sinks may alter the resistance to dislocation movement due to the presence of dissolved or transmuted elements, which accumulate as impurities at grain boundaries, simultaneously triggering radiation hardening processes. The regulation of radiation-induced segregation processes in actual metallic materials are crucial for mitigating radiation damage. This complicated mechanism governing alterations in surface properties may also regulate materials used in fusion reactors, particularly in divertor systems exposed to high-intensity ion effects. The accumulation of radiation defects, phase transitions, and the evolution of the microstructure in metals and alloys are accompanied by alterations in the characteristics of irradiated materials. In the structural materials of atomic and thermonuclear power engineering, the primary issues pertain to radiation hardening, swelling, and alterations in properties and strength. The issues of low-temperature hardening and embrittlement are crucial for the stabilization of superconducting materials in magnetic systems of thermonuclear equipment, as well as for structural materials in space technology. The resolution of these issues is significantly dependent upon the accurate interpretation of radiation damage processes and the advancement of strategies to mitigate the adverse effects of radiation. The degradation mechanisms of metallic materials under irradiation are associated with the distinctive characteristics of structural-phase change mechanisms. Their determination is influenced by the crystal lattice type, chemical and phase composition, dislocation structure, and irradiation parameters, including temperature, radiation type, irradiation energy and intensity. Under the operational conditions of materials in components of nuclear power systems and other systems, the severity and degree of degradation in irradiated materials, as well as their overall span, are significantly influenced by the nature and magnitude of applied stresses, in addition to other factors. This issue is particularly critical for the structural materials of the wall of fusion reactors, which function under the extreme conditions, as well as for the stressed components of nuclear power plants.

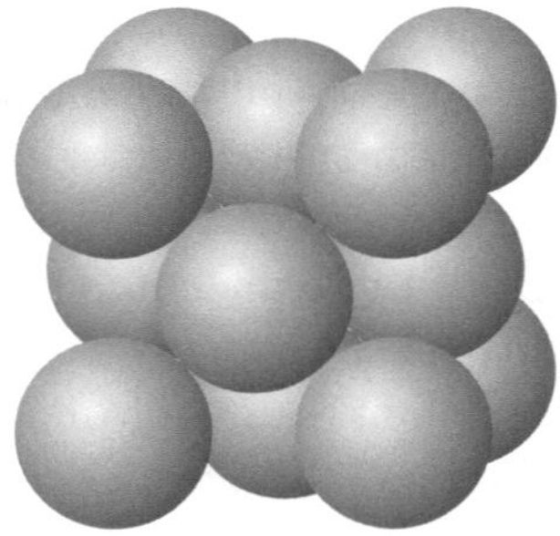

Figure 8.4. *FCC metal.*

Source: https://upload.wikimedia.org/wikipedia/commons/f/fa/FCC_crystal_structure.svg, Creative Commons License.

Figure 8.5. *BCC metal.*

Source: https://upload.wikimedia.org/wikipedia/commons/0/01/Iron_bcc.png, Creative Commons License.

8.9. CRYSTALS

The primary focus of initial investigations into the impact of radiation on the strength and ductility of metals was predominantly conducted on single crystals and polycrystals of copper. The experimental data enable the delineation of the relationships that dictate the fluctuation of the mechanical characteristics of pure FCC (Figure 8.4) and BCC (Figure 8.5) metals (Voyiadjis and Abed,

2005). Some of them are as follows. The enhancement of the mechanical properties of irradiation metals like hardness, tensile strength, yield stress. The yield stress increases intensely than the tensile stress, with a trend to reduce the difference between these parameters at high radiation doses. The reduction of radiation hardening with elevated irradiation or testing temperature is crucial. These relationships that dictate the variation of mechanical characteristics in irradiated metals illustrate the temperature and dose dependency of the interaction between mobile dislocations and radiation defects, which serve as effective barriers to dislocation mobility.

The degradation of mechanical characteristics may be influenced by temperature and irradiation parameters, primarily through the controlling effects of various radiation defects, such as clusters, dislocation loops, or voids, and their combined effects. The fluctuation in defect concentration and size also influences the kinetics of radiation hardening. The dislocation network that may develop during irradiation due to the interaction of radiation-induced defects with the original dislocation structure can serve as an efficient element of radiation hardening. The investigation of temperature influence on radiation hardening in BCC and FCC metals may explain the characteristics of radiation defects as barriers to dislocation motion. In BCC metals, which exhibit a noticeable temperature dependence of yield stress in the non-irradiated state, the mechanism governing hardening post-irradiation remains unchanged. In FCC metals, the low temperature dependence of yield stress in the non-irradiated state intensifies post-irradiation. This signifies that in both BCC and FCC metals, radiation defects predominantly contribute to the thermally activated component of flow stress; specifically, for mobile dislocations, they primarily act as barriers with short-range stress fields. The resolution of these barriers via dislocations occurs due to both applied stresses and thermally stimulated processes. This also establishes the observed temperature dependency of the flow stress in irradiated metals. The theoretically and practically significant correlations of radiation hardening illustrate the dose-dependent saturation of the mechanical properties of exposed metals. The analysis of relevant experimental data and modeling indicates that the kinetics of the process is governed by the mechanism of approaching quasi-equilibrium in the structure of radiation defects, which dictates radiation hardening. The quasi-equilibrium condition persists until the concentration and dimensions of the defects begin to experience substantial alterations due to other processes. In the instance of electron irradiation of vanadium, researchers identified a prominent correlation between the kinetics of microhardness alterations exhibiting a saturation tendency and the kinetics of loop size augmentation.

8.10. DEFECTS

During the experiments, following the application of specific radiation doses, simultaneous measurements of microhardness were conducted, and transmission electron microscopy (Figure 8.6) was employed to ascertain the density and size of the loops, as well as to evaluate the overall concentration of interstitials within the loops. The duration of quasi-equilibrium and the transition period to a new quasi-equilibrium state will initially be dictated by the parameters of radiation-enhanced self-diffusion, specifically the temperature and the rate of annealing of radiation defects, as well as the nature and magnitude of applied stress.

Figure 8.6. *Transmission electron microscopy.*

Source: https://upload.wikimedia.org/wikipedia/commons/1/1f/Ernst_Ruska_Electron_Microscope_-_Deutsches_Museum_-_Munich-edit.jpg, Creative Commons License.

The stress component may significantly influence both the intensification of thermally stimulated diffusion processes related to the growth of the defective structure and the direct interaction of active dislocations with radiation defects. The subsequent transition processes and the quasi-equilibrium condition due to temperature and radiation dose may be regulated by the mechanisms of void generation and radiation-induced recovery of the dislocation structure. Substitutional and interstitial impurities in both BCC and FCC metals, together with alloying elements in diluted alloys, result in enhanced radiation hardening, particularly at moderate radiation doses and temperatures. The non-irradiated alloys exhibit a linear correlation between microhardness and concentration, indicative of the hardening characteristic of diluted solid solutions. The hardening of irradiation alloys exhibits a tendency to saturate with respect to concentration dependency.

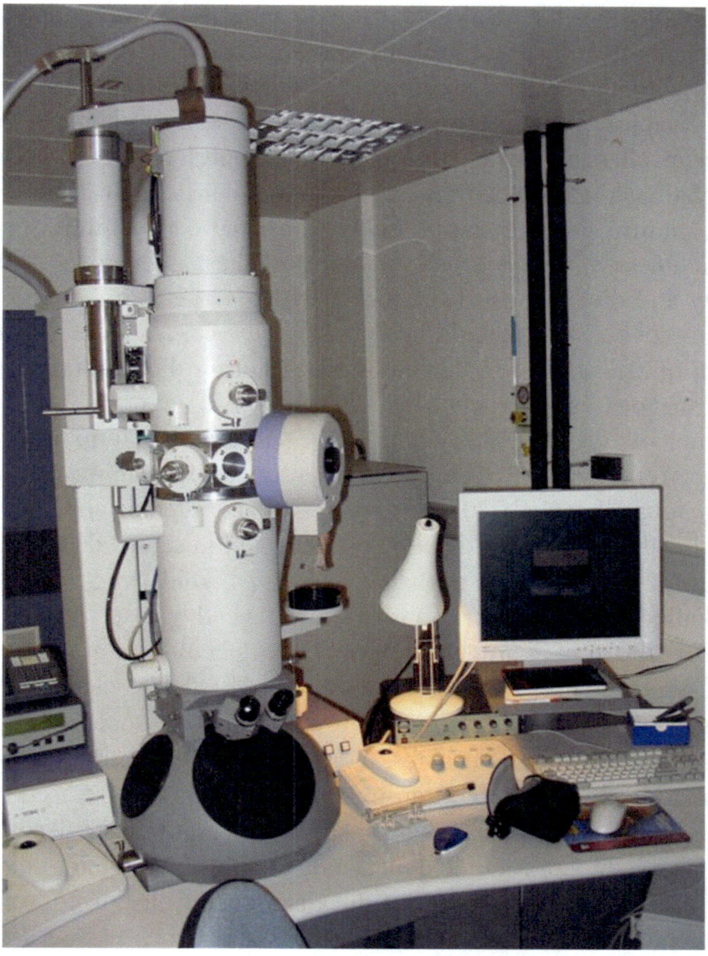

Figure 8.7. *Electron microscope.*

Source: https://upload.wikimedia.org/wikipedia/commons/c/c5/Electron_Microscope.jpg, Creative Commons License.

Following electron microscopic study (Figure 8.7) of irradiated aluminum and aluminum-magnesium alloys indicates that void formation occurs exclusively in pure aluminum during irradiation. In both aluminum and aluminum alloys, analysis revealed dislocation loops, with their concentration rising along with the magnesium content, but their size diminished. So, the concentration dependence of radiation hardening correlates with the density of dislocation loops. The dependence of radiation hardening, reflects the concentration dependence of the nucleation and growth parameters of dislocation loops in aluminum-magnesium alloys of identical composition, subjected to irradiation in a high-voltage electron microscope. Similar to neutron irradiation, the concentration of loops grew with higher magnesium content, whereas the mean size of the loops diminished, approaching saturation with rising concentration. Considering the barrier mechanism of radiation hardening, the electron microscopic data, and the relationships between the density and dimensions of dislocation loops and magnesium content, it can be inferred that the concentration dependence of microhardness variation primarily indicates the tendency for saturation in the density and dimensions of the loops, which serve as effective barriers to dislocation movement. It should be noted that the cascade mechanism for the production of void nuclei in depletion zones during the irradiation of metals with electrons possessing many MeV of energy does not occur. The irradiation of vanadium of identical purity and at the same temperature with electrons possessing an energy of 2.2 MeV resulted in a structure including only interstitial dislocation loops (Lei et al., 2025). More radiation hardening and embrittlement of both diluted undersaturated solid solutions and pure metals, along with an increase in the concentration of dislocation loops, voids, etc., may also result from radiation-stimulated and radiation-enhanced phase transitions. The solubility of silicon in aluminum is low at a temperature of around 55°C. The impact of transmutation silicon on the radiation hardening and embrittlement of aluminum and its alloys is primarily governed by the processes of radiation-induced segregation at various sinks, including grain boundaries, as well as by phase formation. More soluble transmutants may lead to increased hardening due to the dilation distortion of the matrix solid solution or the initial pure metal. The hardening of irradiated metals due to the obstruction of dislocations by radiation-induced defects leads to the emergence of localized high internal stresses, creating conditions conducive to heterogeneous plastic deformation and elevating the likelihood of failure. An instance of the wide localization of plastic yielding, characteristic of both irradiation and FCC, quenched BCC, and hexagonal metals, is the dislocation channeling mechanism. The motion of dislocations is constrained by limited deformation channels, wherein the concentration and magnitude of defects diminish rapidly following dislocation slip. This process is characterized by the development of extensive slip bands and a significant reduction in flexibility. The radiation-accelerated diffusion of these elements to their interfaces leads to the phenomenon of high-temperature radiation embrittlement, demonstrated by a reduction in elongation during both short-term and long-term loading, as well as a preferential failure of the material at these interfaces. In contrast to low-temperature irradiation embrittlement, high-temperature radiation embrittlement is not governed by radiation hardening mechanisms but is influenced by radiation-induced processes occurring at the grain boundaries. A primary factor contributing to the diminished high-temperature plasticity and the exacerbation of intergranular fracture in neutron-irradiated metals is the emergence of clusters and bubbles of transmutation helium at the grain boundaries. In two-phase and multiphase alloys, the influence of phase transitions on mechanical properties, particularly plasticity under specific irradiation conditions, may be significant due to phenomena such as coalescence or the formation of embrittling phases within the matrix, at grain boundaries, and other interfaces.

8.11. ALUMINUM ALLOYS

Aluminum-based alloys are utilized in low-temperature thermal reactors with a water or gas medium for the construction of active zone containers, and other structural components (Czerwinski, 2020). During a certain timeframe, aluminum alloys were considered viable materials for the first wall of fusion reactors, particularly the INTOR reactor (Figure 8.8). This approach was discarded primarily due to two reasons: the shift from the INTOR experimental reactor concept to the ITER reactor, which operates at a higher first wall temperature, and environmental concerns related to the generation of long-lived radionuclides ^{26}Al and ^{22}Na resulting from threshold neutron reactions with energies exceeding 14 MeV. This does not exclude the utilization of aluminum-based alloys as reduced-activation materials beyond the confines of the first wall under conditions of softer neutron spectra. Aluminum alloys continue to be among the promising reduced-activation materials for structural components of thermal neutron reactors. Among industrial aluminum-based alloys, the low-alloyed variants derived from Al-Mg-Si (avials) and Al-Mg-Mn (magnals) have been utilized extensively. The Al-Mg-Mn alloys do not undergo hardening during heat treatment and are utilized in their annealed state. The dose dependencies predominantly represent the correlations of the mechanical property variations previously analyzed for pure metals: radiation hardening and embrittlement, exhibiting a tendency for saturation of these attributes at comparatively elevated radiation doses. Unlike pure metals, the ductility of the irradiated 6061 and 5154 alloys diminishes as test temperatures rise. A potential foundation for this phenomenon may be linked to the extra development of Mg_2Si phase precipitates from the supersaturated solid solution of transmutation silicon in aluminum during heating to the test temperature.

Figure 8.8. *INTOR reactor.*

Source: https://upload.wikimedia.org/wikipedia/commons/7/75/ITER_Exhibit_%2801810402%29_%2812219071813%29_%28cropped%29.jpg, Creative Commons License.

The elevated radiation resistance of the aluminum-magnesium-scandium alloy system is mostly attributable to the following factors: The exceptional efficacy of scandium as an alloying agent in enhancing the mutual recombination of interstitials and vacancies within the solid solution, the stability of the scandium recombination effect within the matrix solid solution due to its minimal susceptibility to radiation-induced segregation, the efficacy of coherent $ScAl_3$ phase precipitates as a neutral sink, facilitating the recombination of vacancies and interstitials, and the stability of the $ScAl_3$ phase particles against radiation-induced coalescence. Besides the considerable radiation resistance of the Al-Mg-Sc system alloys, it is essential to highlight several additional advantageous characteristics, particularly in comparison to the avials, magnals, and alloys from the 6,000 and 5,000 series. The Al-Mg-Sc system alloys are classified as weldable aluminum alloys that are not strengthened through heat treatment. Unlike the alloys of the 6,000 series, these alloys do not necessitate prolonged heat treatment to achieve the desired characteristics. The necessary characteristics are achieved through short-term annealing at temperatures of around 320–420°C. Regarding mechanical properties, these alloys in their initial state surpass those of the 5,000 series, including the 5,154 alloy. In terms of strength, the alloys of the Al-Mg-Sc system are only inferior to several dispersion-hardened alloys in the 6,000 series, particularly alloy 6,061.

8.12. STEEL

Low-alloy ferritic-pearlitic and pearlitic steels are widely utilized in both Russian and international atomic energy systems for pressure vessels in water-water reactors, steam generators, pipelines, and structural components of various functional systems. In Russia, low-alloy ferritic-pearlitic steels of the 22K type were utilized as structural materials for auxiliary equipment, such as steam generator vessels. Similar steels were initially employed directly for the vessels of water-water reactors. This was succeeded by the advancement of steels classified as A553B and A508 in various grades for the fabrication of vessels in industrial power units of nuclear power plants. Compared to steels A212B and A302, they are partially or entirely modified with Mo, Ni, and Cr, with negligible quantities of vanadium (Luce, 1956). These materials are currently employed in vessel structures in the United States, Japan, and France. The chemical composition of these materials, together with the composition of the vessel steels $_{22}NiMoCr_{37}$ and $_{20}MnMoNi_{55}$, produced in Germany, has been documented. They are utilized in the state following quenching and high-temperature tempering, characterized by a tempered bainite structure. The primary materials utilized for the vessels of Russian reactors VVER-440 and VVER-1,000 are heat-resistant steels of the pearlitic type, characterized by a fine-dispersion sorbite microstructure. The operational temperatures of the vessel materials range from around 260 to 320°C, indicating that they function under situations where low-temperature radiation hardening and embrittlement are predominant processes. The primary causes limiting the span of vessel metal to fluxes of about $< 3 \times 10^{20}$ cm^{-2} are a reduction in plasticity and fracture toughness, along with an elevation in the temperature of the ductile–brittle transition. The evaluation of the efficiency of vessel steels during reactor service on irradiated reference specimens have primarily been conducted through measurements of impact toughness characteristics, assessment of fracture toughness in both static and dynamic tests, and the ductile-brittle transition temperature. Russian and international ferritic steels with elevated chromium content from the martensitic or ferritic-martensitic categories are utilized or considered as structural materials for the active zones of atomic reactors. The steels of this grade, along with their reduced-activation Fe-Cr-W compositions, including pearlitic steels with a chromium content of 2–3%, have been studied within several fusion reactor programs. Many studies indicate that the reduced-activation compositions of ferritic steels possess similar technological properties, including radiation resistance, relative to Fe-CrMoVNb and Fe-CrMoV prototypes. There is a significant potential for these materials to replace existing options in both thermonuclear and nuclear power engineering. The majority of experimental research into the mechanical characteristics under active tensile and impact loading have been conducted on ferritic steels irradiated at temperatures ranging from around 300 to 600°C. This temperature range aligns with the actual or intended operational temperatures for the specified material within the context of anticipated industrial applications in nuclear power engineering. Within the studied temperature range, an elevation in irradiation temperature diminishes the degree of radiation hardening in ferritic steels, but the flexibility of the steels progressively increases. This trend is also seen at irradiation temperatures of around 420–450°C, where radiation softening prevails, potentially resulting in a reduction of strength characteristics below those of the initial, non-irradiated state. During the softening stage, the concentration of dislocation-type radiation defects significantly diminishes, and the overall initial microstructure undergoes minimal alteration. Following irradiation at 600°C, the alterations in the microstructure and mechanical properties of martensitic steels closely resemble the changes found following prolonged thermal ageing at the same temperature. Softening similar to ageing is linked to the reconfiguration of martensitic lath borders and the enlargement of subgrains and precipitates, attributable to radiation-enhanced diffusion processes. Similar to the irradiation of pure metals and vessel steels, ferritic chromium steels exhibit saturation of radiation hardening at increased temperatures, occurring at significantly higher temperatures and within the range of 390–400°C. In pure metals and ferritic steels, saturation of radiation hardening and embrittlement is observed at fluences at 0.15 dpa. At temperatures ranging from 390 to 430°C, ferritic steels containing 9 and 12% Cr can maintain a quasi-equilibrium state at elevated fluences, equating to around 70 or even 100 dpa. At equivalent fluences, the radiation hardening at elevated temperatures during irradiation is typically maximized for ferritic steels containing 12% Cr (Luce, 1956). In low-alloy ferritic steels, functioning at comparatively low temperatures, radiation hardening and embrittlement, similar to pure metals, are primarily governed by the configuration of radiation defects. In pearlitic, martensitic, and ferritic–martensitic steels irradiated at elevated temperatures, the strength and ductility parameters are significantly influenced by radiation-induced phase transformations, along with the radiation defects. The saturation of the temperature shift of the ductile-brittle transition in both reduced-activation ferritic steels and their prototypes containing Mo and Nb is often observed at fluences ranging from approximately 10 to 20 dpa. This outcome correlates with the saturation of radiation hardening and embrittlement in irradiated ferritic steels at elevated temperatures. The mechanical properties of ferritic steels, those containing 9% Cr and subjected to extreme temperatures, conform with the criteria for their use as structural materials in fast reactors. They are considered potential materials for the first wall of fusion reactors. A distinctive feature of structural materials in fusion reactors is the accelerated transmutation of gaseous products from nuclear reactions, particularly helium and hydrogen. The primary challenges are related to helium formation. The influence of helium on radiation hardening and embrittlement of ferritic steels at extreme temperatures, in comparison to austenitic steels, is relatively weak and does not pose a limiting issue for potential applications in thermonuclear power engineering. The elevated concentrations lead to significant low-temperature embrittlement of the irradiation ferritic steels. The irradiation of steels in the HFIR reactor, utilizing a mixed neutron spectrum at temperatures ranging from 300 to 400°C, results in atomic displacements and helium concentrations, hence elevating their ductile-brittle transition temperature by 200–240°C. An additional increase in helium concentration to 2×10^{-2} due to 2% Ni shifts their transition temperature by approximately 350°C following irradiation. Following irradiation of identical materials in an intense neutron spectrum of the fast EBRII reactor at a temperature of 390°C, the transition temperature shift becomes saturated, with a fluence of 26 dpa resulting in shifts of 144°C and 54°C for steels containing 12% and 9% Cr, respectively. Austenitic chromium-nickel steels and their analogues and modifications, are extensively utilized or considered promising structural materials for pipelines and heat exchangers in water-water reactors, as well as for vessels, and assemblies of fuel elements in fast neutron reactors employing liquid metal heat carriers. These materials, together with reduced-activation austenitic chromium–manganese steels are considered potential structural materials for fusion reactors. For hexagonal wrapper tubes of fuel components and fuel assembly, it is advisable to employ cold plastic deformation following austenitizing to mitigate swelling. The viability of such an operation may also be important regarding the potential use of austenitic steels as structural materials for fusion reactors. The HTRE impact of these materials in neutron radiation is primarily influenced by the melting technology, chemical composition, and/or the mechanisms of radiation-induced segregation of dissolved components at the grain boundaries. The phase transformations in chromium–nickel austenitic steels under irradiation have been thoroughly analyzed. The primary influence on the evolution of the microstructure and mechanical characteristics of irradiated steels is derived from radiation-stimulated and radiation-induced processes of segregation, matrix solution decomposition, and coalescence.

BIBLIOGRAPHY

1. Abdelaleem, M. A. (2017). Radionuclides in foods. In *Food Toxicology*. Apple Academic Press.
2. Abuelhia, E., Kacperski, K., & Spyrou, N. M. (2007). Three-photon annihilation in PET: 2D imaging experiments. *Journal of Radioanalytical and Nuclear Chemistry, 271*(2), 489–495. https://doi.org/10.1007/s10967-007-0235-9 (accessed on 30th July 2025).
3. Allen, T. R., & Was, G. S. (2007). Radiation-enhanced diffusion and radiation-induced segregation. In K. E. Sickafus, E. A. Kotomin, & B. P. Uberuaga (Eds.), *Radiation Effects in Solids* (Vol. 235, pp. 123–151). Springer Netherlands. https://doi.org/10.1007/978-1-4020-5295-8_6 (accessed on 30th July 2025).
4. Alqahtani, F. F. (2023). SPECT/CT and PET/CT, related radiopharmaceuticals, and areas of application and comparison. *Saudi Pharmaceutical Journal, 31*(2), 312–328. https://doi.org/10.1016/j.jsps.2022.12.013 (accessed on 30th July 2025).
5. Alvarenga, M. A. B., & Frutuoso e Melo, P. F. (2015). Including severe accidents in the design basis of nuclear power plants: An organizational factors perspective after the Fukushima accident. *Annals of Nuclear Energy, 79*, 68–77. https://doi.org/10.1016/j.anucene.2015.01.016 (accessed on 30th July 2025).
6. Apte, K., & Bhide, S. (2024). Chapter 1—Basics of radiation. In S. Verma & A. K. Srivastava (Eds.), *Advanced Radiation Shielding Materials* (pp. 1–23). Elsevier. https://doi.org/10.1016/B978-0-323-95387-0.00013-3 (accessed on 30th July 2025).
7. Arbutina, D., Vasić-Milovanović, A., & Nedić, T. (2022). Statistical methods for designing a Geiger–Muller counter suitable for an early

warning system. *Nuclear Instruments and Methods in Physics Research Section A: Accelerators, Spectrometers, Detectors and Associated Equipment, 1040*, 167268. https://doi.org/10.1016/j.nima.2022.167268 (accessed on 30th July 2025).

8. Ariola, M., & Pironti, A. (2016). *Magnetic Control of Tokamak Plasmas*. Springer International Publishing. https://doi.org/10.1007/978-3-319-29890-0 (accessed on 30th July 2025).
9. Babenko, V. A., Jenkovszky, L. L., & Pavlovych, V. N. (2007). Nuclear power industry: Tendencies in the world and Ukraine. *Physics of Particles and Nuclei, 38*(6), 795–826. https://doi.org/10.1134/S1063779607060056 (accessed on 30th July 2025).
10. Bair, J. K., Miller, P. D., & Wieland, B. W. (1981). Neutron yields from the 4–12 MeV proton bombardment of 11B, 13C and 18O as related to the production of 11C, 13N and 18F. *The International Journal of Applied Radiation and Isotopes, 32*(6), 389–395. https://doi.org/10.1016/S0020-708X(81)81005-8 (accessed on 30th July 2025).
11. Bastiaans, G. J., & Mangold, R. A. (1985). The calculation of electron density and temperature in Ar spectroscopic plasmas from continuum and line spectra. *Spectrochimica Acta Part B: Atomic Spectroscopy, 40*(7), 885–892. https://doi.org/10.1016/0584-8547(85)80059-8 (accessed on 30th July 2025).
12. Baur, G., Hencken, K., & Trautmann, D. (2007). Electron–positron pair production in ultra relativistic heavy ion collisions. *Physics Reports, 453*(1), 1–27. https://doi.org/10.1016/j.physrep.2007.09.002 (accessed on 30th July 2025).
13. Bawazeer, O., Makkawi, K., Aga, Z. B., Albakri, H., Assiri, N., Althagafy, K., & Ajlouni, A. W. (2023). A review on using nanocomposites as shielding materials against ionizing radiation. *Journal of Umm Al-Qura University for Applied Sciences, 9*(3), 325–340. https://doi.org/10.1007/s43994-023-00042-9 (accessed on 30th July 2025).
14. Bellon, P. (2009). Precipitate and microstructural stability in alloys subjected to sustained irradiation. In H. Bernas (Ed.), *Materials Science with Ion Beams* (Vol. 116, pp. 29–52). Springer Berlin Heidelberg. https://doi.org/10.1007/978-3-540-88789-8_2 (accessed on 30th July 2025).
15. Belyaev, A., & Ross, D. (2021). Radioactivity. In A. Belyaev & D. Ross (Eds.), *The Basics of Nuclear and Particle Physics* (pp. 67–88). Springer International Publishing. https://doi.org/10.1007/978-3-030-80116-8_5 (accessed on 30th July 2025).
16. Bhabha, H. J., Heitler, W., & Mott, N. F. (1997). The passage of fast

electrons and the theory of cosmic showers. *Proceedings of the Royal Society of London. Series A–Mathematical and Physical Sciences, 159*(898), 432–458. https://doi.org/10.1098/rspa.1937.0082 (accessed on 30th July 2025).

17. Bhatia, C., Fallin, B. F., Gooden, M. E., Howell, C. R., Kelley, J. H., Tornow, W., Arnold, C. W., Bond, E., Bredeweg, T. A., Fowler, M. M., Moody, W., Rundberg, R. S., Rusev, G. Y., Vieira, D. J., Wilhelmy, J. B., Becker, J. A., Macri, R., Ryan, C., Sheets, S. A., & Tonchev, A. P. (2015). Exploratory study of fission product yields of neutron-induced fission of ^{235}U, ^{238}U, and ^{239}Pu at 8.9 MeV. *Physical Review C, 91*(6), 064604. https://doi.org/10.1103/PhysRevC.91.064604 (accessed on 30th July 2025).
18. Bockisch, A., Freudenberg, L. S., Schmidt, D., & Kuwert, T. (2009). Hybrid imaging by SPECT/CT and PET/CT: Proven outcomes in cancer imaging. *Seminars in Nuclear Medicine, 39*(4), 276–289. https://doi.org/10.1053/j.semnuclmed.2009.03.003 (accessed on 30th July 2025).
19. Boulos, M. I., Fauchais, P. L., & Pfender, E. (2015). Plasma radiation transport. In *Handbook of Thermal Plasmas* (pp. 1–74). Springer, Cham. https://doi.org/10.1007/978-3-319-12183-3_8-1 (accessed on 30th July 2025).
20. Burgess, A., & Percival, I. C. (1968). Classical theory of atomic scattering. In D. R. Bates & I. Estermann (Eds.), *Advances in Atomic and Molecular Physics* (Vol. 4, pp. 109–141). Academic Press. https://doi.org/10.1016/S0065-2199(08)60186-4 (accessed on 30th July 2025).
21. Calabrese, E. J. (2022). Linear non-threshold (LNT) fails numerous toxicological stress tests: Implications for continued policy use. *Chemico-Biological Interactions, 365*, 110064. https://doi.org/10.1016/j.cbi.2022.110064 (accessed on 30th July 2025).
22. Camacho-Olguín, C. A., Garcia-Borquez, A., Paz Del Angel, V., Fuente, J. A. M. D. L., González-Rodríguez, C. A., Ramos-Tercero, L. M., Cruz-Mejia, H., & Rodriguez-Diaz, R. A. (2020). Formation of the $Ni_{31}Si_{12}$ phase induced by the irradiation of nickel ions on the eutectic, α-Ni-Ni_3Si-monoclinic, with a 380 dpa dose. *Radiation Effects and Defects in Solids, 175*(9, 10), 925–937. https://doi.org/10.1080/10420150.2020.1793338 (accessed on 30th July 2025).
23. Cerrito, L. (2017). Natural sources of radiation. In L. Cerrito (Ed.), *Radiation and Detectors: Introduction to the Physics of Radiation and Detection Devices* (pp. 19–36). Springer International Publishing. https://doi.org/10.1007/978-3-319-53181-6_2 (accessed on 30th July 2025).
24. Cesareo, R., Hanson, A. L., Gigante, G. E., Pedraza, L. J., & Mathaboally, S.

Q. G. (1992). Interaction of keV photons with matter and new applications. *Physics Reports, 213*(3), 117–178. https://doi.org/10.1016/0370-1573(92)90086-F (accessed on 30th July 2025).

25. Chizhik, V. I., Chernyshev, Y. S., Donets, A. V., Frolov, V. V., Komolkin, A. V., & Shelyapina, M. G. (2014). *Magnetic Resonance and Its Applications*. Springer International Publishing. https://doi.org/10.1007/978-3-319-05299-1 (accessed on 30th July 2025).
26. Clauser, C. (2024). Radioactive decay and rock ages. In C. Clauser (Ed.), *Introduction to Geophysics: Global Physical Fields and Processes in the Earth* (pp. 31–53). Springer International Publishing. https://doi.org/10.1007/978-3-031-17867-2_2 (accessed on 30th July 2025).
27. Clough, R. L. (2001). High-energy radiation and polymers: A review of commercial processes and emerging applications. *Nuclear Instruments and Methods in Physics Research Section B: Beam Interactions with Materials and Atoms, 185*(1), 8–33. https://doi.org/10.1016/S0168-583X(01)00966-1 (accessed on 30th July 2025).
28. Cook, G. T., & MacKenzie, A. B. (2014). Radioactive isotope analyses of skeletal materials in forensic science: A review of uses and potential uses. *International Journal of Legal Medicine, 128*(4), 685–698. https://doi.org/10.1007/s00414-014-0970-8 (accessed on 30th July 2025).
29. Cook, S., D'Arcy, R., Edmonds, A., Fukuda, M., Hatanaka, K., Hino, Y., Kuno, Y., Lancaster, M., Mori, Y., Ogitsu, T., Sakamoto, H., Sato, A., Tran, N. H., Truong, N. M., Wing, M., Yamamoto, A., & Yoshida, M. (2017). Delivering the world's most intense muon beam. *Physical Review Accelerators and Beams, 20*(3), 030101. https://doi.org/10.1103/PhysRevAccelBeams.20.030101 (accessed on 30th July 2025).
30. Correia, J. H., Rodrigues, J. A., Pimenta, S., Dong, T., & Yang, Z. (2021). Photodynamic therapy review: Principles, photosensitizers, applications, and future directions. *Pharmaceutics, 13*(9), Article 9. https://doi.org/10.3390/pharmaceutics13091332 (accessed on 30th July 2025).
31. Coursey, B. M., & Nath, R. (2000). Radionuclide therapy. *Physics Today, 53*(4), 25–30. https://doi.org/10.1063/1.883033 (accessed on 30th July 2025).
32. Czerwinski, F. (2020). Thermal stability of aluminum alloys. *Materials, 13*(15), 3441. https://doi.org/10.3390/ma13153441 (accessed on 30th July 2025).
33. D'Auria, S. (2018). Radioactive decays. In S. D'Auria (Ed.), *Introduction to Nuclear and Particle Physics* (pp. 51–70). Springer International

Publishing. https://doi.org/10.1007/978-3-319-93855-4_4 (accessed on 30th July 2025).

34. Darken, L. S., & Cox, C. E. (1995). Chapter 2 High-purity germanium detectors. In *Semiconductors and semimetals* (Vol. 43, pp. 23–83). Elsevier. https://doi.org/10.1016/S0080-8784(08)62741-6 (accessed on 30th July 2025).

35. De Los Santos, J., Popple, R., Agazaryan, N., Bayouth, J. E., Bissonnette, J. P., Bucci, M. K., Dieterich, S., Dong, L., Forster, K. M., Indelicato, D., Langen, K., Lehmann, J., Mayr, N., Parsai, I., Salter, W., Tomblyn, M., Yuh, W. T. C., & Chetty, I. J. (2013). Image guided radiation therapy (IGRT) technologies for radiation therapy localization and delivery. *International Journal of Radiation Oncology Biology Physics, 87*(1), 33–45. https://doi.org/10.1016/j.ijrobp.2013.02.021 (accessed on 30th July 2025).

36. Dörner, R., Weber, Th., Weckenbrock, M., Staudte, A., Hattass, M., Schmidt-Böcking, H., Moshammer, R., & Ullrich, J. (2002). Multiple ionization in strong laser fields. In B. Bederson & H. Walther (Eds.), *Advances in Atomic, Molecular, and Optical Physics* (Vol. 48, pp. 1–34). Academic Press. https://doi.org/10.1016/S1049-250X(02)80005-2 (accessed on 30th July 2025).

37. Dumazert, J., Coulon, R., Lecomte, Q., Bertrand, G. H. V., & Hamel, M. (2018). Gadolinium for neutron detection in current nuclear instrumentation research: A review. *Nuclear Instruments and Methods in Physics Research Section A: Accelerators, Spectrometers, Detectors and Associated Equipment, 882*, 53–68. https://doi.org/10.1016/j.nima.2017.11.032 (accessed on 30th July 2025).

38. Elgazzar, A. H., & Sarikaya, I. (2022). Basis of therapeutic nuclear medicine. In A. H. Elgazzar (Ed.), *The Pathophysiologic Basis of Nuclear Medicine* (pp. 569–594). Springer International Publishing. https://doi.org/10.1007/978-3-030-96252-4_13 (accessed on 30th July 2025).

39. Enger, S. A., Ahnesjo, A., & Beaulieu, L. (2010). Dose to medium or dose to a water cavity embedded in medium? *Brachytherapy, 9*, S35. https://doi.org/10.1016/j.brachy.2010.02.036 (accessed on 30th July 2025).

40. Evans, J. A., DeHart, M. D., Weaver, K. D., & Keiser, D. D. (2022). Burnable absorbers in nuclear reactors–A review. *Nuclear Engineering and Design, 391*, 111726. https://doi.org/10.1016/j.nucengdes.2022.111726 (accessed on 30th July 2025).

41. Fagerstrom, J. M. (2020). Computed tomography, sinograms, and image reconstruction in the classroom. *Physics Education, 55*(3), 034001. https://doi.org/10.1088/1361-6552/ab753c (accessed on 30th July 2025).

42. Fantz, U., Briefi, S., Heiler, A., Wimmer, C., & Wünderlich, D. (2021). Negative hydrogen ion sources for fusion: From plasma generation to beam properties. *Frontiers in Physics, 9*. https://doi.org/10.3389/fphy.2021.709651 (accessed on 30th July 2025).

43. Filburn, T., & Bullard, S. (2016a). Three Mile Island. In T. Filburn & S. Bullard, *Three Mile Island, Chernobyl and Fukushima* (pp. 29–41). Springer International Publishing. https://doi.org/10.1007/978-3-319-34055-5_3 (accessed on 30th July 2025).

44. Filburn, T., & Bullard, S. (2016b). *Three Mile Island, Chernobyl and Fukushima: Curse of the nuclear genie*. Springer International Publishing. https://doi.org/10.1007/978-3-319-34055-5 (accessed on 30th July 2025).

45. Forty, R., & Ullaland, O. (2020). Particle identification: Time-of-flight, Cherenkov and transition radiation detectors. In C. W. Fabjan & H. Schopper (Eds.), *Particle Physics Reference Library: Volume 2: Detectors for Particles and Radiation* (pp. 281–335). Springer International Publishing. https://doi.org/10.1007/978-3-030-35318-6_7 (accessed on 30th July 2025).

46. Freed, K. F. (2002). Radiation less transitions in molecules (world) [Research-article]. *ACS Publications*; American Chemical Society. https://doi.org/10.1021/ar50122a004 (accessed on 30th July 2025).

47. Fynan, D. A., & Park, J. (2021). Operator action–induced two-phase flow condition resulting in performance degradation of interfacing passive system. *Nuclear Technology, 207*(3), 335–351. https://doi.org/10.1080/00295450.2020.1760704 (accessed on 30th July 2025).

48. Garashchenko, B. L., Korsakova, V. A., & Yu Yakovlev, R. (2018). Radiopharmaceuticals based on alpha emitters: Preparation, properties, and application. *Physics of Atomic Nuclei, 81*(10), 1515–1525. https://doi.org/10.1134/S1063778818100071 (accessed on 30th July 2025).

49. Gauthier-Lafaye, F., Holliger, P., & Blanc, P. L. (1996). Natural fission reactors in the Franceville Basin, Gabon: A review of the conditions and results of a "critical event" in a geologic system. *Geochimica et Cosmochimica Acta, 60*(23), 4831–4852. https://doi.org/10.1016/S0016-7037(96)00245-1 (accessed on 30th July 2025).

50. Golovko, V. V. (2022). Simplified efficiency calibration methods for semiconductor detectors used in criticality dosimetry. *Applied Radiation and Isotopes, 187*, 110335. https://doi.org/10.1016/j.apradiso.2022.110335 (accessed on 30th July 2025).

51. Goncharov, V. V. (1964). Review of work on research reactors and their application in the USSR. *Soviet Atomic Energy, 17*(4), 989–999. https://

doi.org/10.1007/BF01116285 (accessed on 30th July 2025).

52. Greene, D., & Williams, P. C. (2017). *Linear Accelerators for Radiation Therapy* (2nd ed.). CRC Press. https://doi.org/10.1201/9780429246562 (accessed on 30th July 2025).

53. Hachenberg, O., & Brauer, W. (1959). Secondary electron emission from solids. In L. Marton & C. Marton (Eds.), *Advances in Electronics and Electron Physics* (Vol. 11, pp. 413–499). Academic Press. https://doi.org/10.1016/S0065-2539(08)60999-3 (accessed on 30th July 2025).

54. Hänscheid, H., Lassmann, M., & Verburg, F. A. (2023). Determinants of target absorbed dose in radionuclide therapy. *Zeitschrift Für Medizinische Physik, 33*(1), 82–90. https://doi.org/10.1016/j.zemedi.2022.10.001 (accessed on 30th July 2025).

55. Harley, N. H. (2020). Radon and lung cancer. In *Environmental Toxicants* (pp. 877–909). John Wiley & Sons, Ltd. https://doi.org/10.1002/9781119438922.ch23 (accessed on 30th July 2025).

56. Hassfjell, S., & Brechbiel, M. W. (2001). The development of the α-particle emitting radionuclides ^{212}Bi and ^{213}Bi, and their decay chain related radionuclides, for therapeutic applications. *Chemical Reviews, 101*(7), 2019–2036. https://doi.org/10.1021/cr000118y (accessed on 30th July 2025).

57. Hemsing, E., Stupakov, G., Xiang, D., & Zholents, A. (2014). Beam by design: Laser manipulation of electrons in modern accelerators. *Reviews of Modern Physics, 86*(3), 897–941. https://doi.org/10.1103/RevModPhys.86.897 (accessed on 30th July 2025).

58. Hentschel, K. (2018). Twelve semantic layers of 'light quantum' and 'photon.' In K. Hentschel (Ed.), *Photons: The History and Mental Models of Light Quanta* (pp. 39–92). Springer International Publishing. https://doi.org/10.1007/978-3-319-95252-9_3 (accessed on 30th July 2025).

59. Hillier, D. J. (2023). Photoionization and electron–ion recombination in astrophysical plasmas. *Atoms, 11*(3), Article 3. https://doi.org/10.3390/atoms11030054 (accessed on 30th July 2025).

60. Hori, M., & Walz, J. (2013). Physics at CERN's antiproton decelerator. *Progress in Particle and Nuclear Physics, 72*, 206–253. https://doi.org/10.1016/j.ppnp.2013.02.004 (accessed on 30th July 2025).

61. Hsu, H. T., Chen, C. H., & Lo, C. K. (2020). Risk reduction assessment for installing RCP shutdown seal. *Nuclear Technology, 206*(12), 1891–1908. https://doi.org/10.1080/00295450.2020.1731404 (accessed on 30th July 2025).

62. Israel, O., Pellet, O., Biassoni, L., De Palma, D., Estrada-Lobato, E., Gnanasegaran, G., Kuwert, T., La Fougère, C., Mariani, G., Massalha, S., Paez, D., & Giammarile, F. (2019). Two decades of SPECT/CT–the coming of age of a technology: An updated review of literature evidence. *European Journal of Nuclear Medicine and Molecular Imaging, 46*(10), 1990–2012. https://doi.org/10.1007/s00259-019-04404-6 (accessed on 30th July 2025).

63. Jain, M., & Tzoar, N. (1978). Compton scattering in the presence of coherent electromagnetic radiation. *Physical Review A, 18*(2), 538–545. https://doi.org/10.1103/PhysRevA.18.538 (accessed on 30th July 2025).

64. Jarrett, R. D. (1982). Isotope (gamma) radiation sources. In *Preservation of food by ionizing radiation*. CRC Press.

65. Jiang, W., Huang, Y., An, Y., & Kim, B. Y. S. (2015). Remodeling tumor vasculature to enhance delivery of intermediate-sized nanoparticles. *ACS Nano, 9*(9), 8689–8696. https://doi.org/10.1021/acsnano.5b02028 (accessed on 30th July 2025).

66. Johnson, N. S., Mishra, A. A., Kirsch, D. J., & Mehta, A. (2024). Active learning for rapid targeted synthesis of compositionally complex alloys. *Materials, 17*(16), Article 16. https://doi.org/10.3390/ma17164038 (accessed on 30th July 2025).

67. Juhn, P. E., Kupitz, J., Cleveland, J., Cho, B., & Lyon, R. B. (2000). IAEA activities on passive safety systems and overview of international development. *Nuclear Engineering and Design, 201*(1), 41–59. https://doi.org/10.1016/S0029-5493(00)00260-0 (accessed on 30th July 2025).

68. Kallappa, P., Holmes, M. S., & Ray, A. (1997). Life-extending control of fossil fuel power plants. *Automatica, 33*(6), 1101–1118. https://doi.org/10.1016/S0005-1098(97)00014-9 (accessed on 30th July 2025).

69. Karger, C. P., Glowa, C., Peschke, P., & Kraft-Weyrather, W. (2021). The RBE in ion beam radiotherapy: *In vivo* studies and clinical application. *Zeitschrift Für Medizinische Physik, 31*(2), 105–121. https://doi.org/10.1016/j.zemedi.2020.12.001 (accessed on 30th July 2025).

70. Khalil, M. M. (2021). Internal radiation dosimetry. In M. M. Khalil (Ed.), *Basic Sciences of Nuclear Medicine* (pp. 203–228). Springer International Publishing. https://doi.org/10.1007/978-3-030-65245-6_9 (accessed on 30th July 2025).

71. Khene, S. (2021). Black body. In S. Khene (Ed.), *Topics and Solved Exercises at the Boundary of Classical and Modern Physics* (pp. 231–284). Springer International Publishing. https://doi.org/10.1007/978-3-030-87742-2_3 (accessed on 30th July 2025).

72. Kim, S. H., Chang, S. H., Choi, Y. J., & Jeong, Y. H. (2015). A passive decay heat removal strategy of the integrated passive safety system (IPSS) for SBO combined with LOCA. *Nuclear Engineering and Design, 295*, 346–359. https://doi.org/10.1016/j.nucengdes.2015.09.033 (accessed on 30th July 2025).

73. Korol, A. V., & Solov'yov, A. V. (2014). *Polarization Bremsstrahlung* (Vol. 80). Springer. https://doi.org/10.1007/978-3-642-45224-6 (accessed on 30th July 2025).

74. Kozlovski, V. V., Vasil'ev, A. E., & Lebedev, A. A. (2015). Role of the recoil atom energy in the formation of radiation-induced defects in semiconductors under electron bombardment. *Journal of Surface Investigation. X-Ray, Synchrotron and Neutron Techniques, 9*(2), 231–236. https://doi.org/10.1134/S1027451015020123 (accessed on 30th July 2025).

75. Kraft, G. (2000). Tumor therapy with ion beams. *Nuclear Instruments and Methods in Physics Research Section A: Accelerators, Spectrometers, Detectors and Associated Equipment, 454*(1), 1–10. https://doi.org/10.1016/S0168-9002(00)00802-0 (accessed on 30th July 2025).

76. Krishnamoorthy, S., Schmall, J. P., & Surti, S. (2017). PET physics and instrumentation. In M. M. Khalil (Ed.), *Basic Science of PET Imaging* (pp. 173–197). Springer International Publishing. https://doi.org/10.1007/978-3-319-40070-9_8 (accessed on 30th July 2025).

77. Langel, W. (2023). Introduction to neutron scattering. *ChemTexts, 9*(4), 12. https://doi.org/10.1007/s40828-023-00184-7 (accessed on 30th July 2025).

78. Lea, K. R. (1978). Highlights of synchrotron radiation. *Physics Reports, 43*(8), 337–375. https://doi.org/10.1016/0370-1573(78)90202-8 (accessed on 30th July 2025).

79. Lei, T., Hu, C., Zhang, Q., & Wang, X. (2025). Elemental segregation and solute effects on mechanical properties and processing of vanadium alloys: A review. *Metals, 15*(1), 96. https://doi.org/10.3390/met15010096 (accessed on 30th July 2025).

80. Lépy, M. C., Amiot, M. N., Bé, M. M., & Cassette, P. (2006). Determination of the intensity of X- and gamma-ray emissions in the decay of ^{153}Sm. *Applied Radiation and Isotopes, 64*(10), 1428–1434. https://doi.org/10.1016/j.apradiso.2006.02.056 (accessed on 30th July 2025).

81. Li, G., Blake, G. R., & Palstra, T. T. M. (2017). Vacancies in functional materials for clean energy storage and harvesting: The perfect imperfection. *Chemical Society Reviews, 46*(6), 1693–1706. https://doi.org/10.1039/

C6CS00571C (accessed on 30th July 2025).

82. Li, W., & Hudson, M. K. (2019). Earth's Van Allen radiation belts: From discovery to the Van Allen Probes era. *Journal of Geophysical Research: Space Physics, 124*(11), 8319–8351. https://doi.org/10.1029/2018JA025940 (accessed on 30th July 2025).
83. Livieratos, L. (2012). Basic principles of SPECT and PET imaging. In I. Fogelman, G. Gnanasegaran, & H. Van Der Wall (Eds.), *Radionuclide and Hybrid Bone Imaging* (pp. 345–359). Springer Berlin Heidelberg. https://doi.org/10.1007/978-3-642-02400-9_12 (accessed on 30th July 2025).
84. Livingston, M. S. (1952). High energy accelerators: Standard cyclotron. *Annual Review of Nuclear and Particle Science, 1*(Volume 1, 1952), 157–167. https://doi.org/10.1146/annurev.ns.01.120152.001105 (accessed on 30th July 2025).
85. Luce, W. A. (1956). Stainless steels including other ferrous alloys. *Industrial & Engineering Chemistry, 48*(9), 1775–1787. https://doi.org/10.1021/ie51401a021 (accessed on 30th July 2025).
86. Maddalena, F., Tjahjana, L., Xie, A., Arramel, Zeng, S., Wang, H., Coquet, P., Drozdowski, W., Dujardin, C., Dang, C., & Birowosuto, M. (2019). Inorganic, organic, and perovskite halides with nanotechnology for high–light yield X- and γ-ray scintillators. *Crystals, 9*(2), 88. https://doi.org/10.3390/cryst9020088 (accessed on 30th July 2025).
87. Manno, V. P., & Golay, M. W. (1985). Nuclear power plant design innovation through simplification. *Nuclear Engineering and Design, 85*(3), 315–325. https://doi.org/10.1016/0029-5493(85)90231-6 (accessed on 30th July 2025).
88. Martin, A., & Harbison, S. A. (1980). X-rays and radiography. In A. Martin & S. A. Harbison (Eds.), *An Introduction to Radiation Protection* (pp. 165–175). Springer US. https://doi.org/10.1007/978-1-4899-6940-8_12 (accessed on 30th July 2025).
89. McQuaid, S. J., & Hutton, B. F. (2008). Sources of attenuation-correction artefacts in cardiac PET/CT and SPECT/CT. *European Journal of Nuclear Medicine and Molecular Imaging, 35*(6), 1117–1123. https://doi.org/10.1007/s00259-008-0718-0 (accessed on 30th July 2025).
90. Mehedinteanu, S. (2009). The evaluation of RCS depressurization to prevent core melting in pressure tube reactors (CANDU-type). *Annals of Nuclear Energy, 36*(4), 537–545. https://doi.org/10.1016/j.anucene.2008.11.021 (accessed on 30th July 2025).
91. Mehlhorn, W. (1983). Electron decay processes. In H. O. Lutz, J. S. Briggs, & H. Kleinpoppen (Eds.), *Fundamental Processes in Energetic Atomic*

Collisions (pp. 253–293). Springer US. https://doi.org/10.1007/978-1-4613-3781-2_7 (accessed on 30th July 2025).

92. Melcher, C. L., Eriksson, L. A., Aykac, M., Bauer, F., Williams, C., Loope, M., & Schmand, M. (2006). Current and future use of LSO:Ce scintillators in PET. In S. Tavernier, A. Gektin, B. Grinyov, & W. W. Moses (Eds.), *Radiation Detectors for Medical Applications* (pp. 243–257). Springer Netherlands. https://doi.org/10.1007/1-4020-5093-3_10 (accessed on 30th July 2025).
93. Meyzonnette, J. L., Mangin, J., & Cathelinaud, M. (2019). Refractive index of optical materials. In J. D. Musgraves, J. Hu, & L. Calvez (Eds.), *Springer Handbook of Glass* (pp. 997–1045). Springer International Publishing. https://doi.org/10.1007/978-3-319-93728-1_29 (accessed on 30th July 2025).
94. Michaelides (Stathis), E. E. (2012). Nuclear power plants. In E. E. (Stathis) Michaelides (Ed.), *Alternative Energy Sources* (pp. 131–172). Springer. https://doi.org/10.1007/978-3-642-20951-2_5 (accessed on 30th July 2025).
95. Morgenstern, A., Apostolidis, C., & Bruchertseifer, F. (2020). Supply and clinical application of actinium-225 and bismuth-213. *Seminars in Nuclear Medicine, 50*(2), 119–123. https://doi.org/10.1053/j.semnuclmed.2020.02.003 (accessed on 30th July 2025).
96. Müller, A. (1991). Ion formation processes: Ionization in ion-electron collisions. In D. Mathur (Ed.), *Physics of Ion Impact Phenomena* (pp. 13–90). Springer. https://doi.org/10.1007/978-3-642-84350-1_2 (accessed on 30th July 2025).
97. Nag, S., Chao, C., Erickson, B., Fowler, J., Gupta, N., Martinez, A., & Thomadsen, B. (2002). The American Brachytherapy Society recommendations for low-dose-rate brachytherapy for carcinoma of the cervix. *International Journal of Radiation Oncology Biology Physics, 52*(1), 33–48. https://doi.org/10.1016/S0360-3016(01)01755-2 (accessed on 30th July 2025).
98. Nag, S., Erickson, B., Thomadsen, B., Orton, C., Demanes, J. D., & Petereit, D. (2000). The American Brachytherapy Society recommendations for high-dose-rate brachytherapy for carcinoma of the cervix. *International Journal of Radiation Oncology Biology Physics, 48*(1), 201–211. https://doi.org/10.1016/S0360-3016(00)00497-1 (accessed on 30th July 2025).
99. Naik, P. A., Arora, V., Prasad, Y. B. S. R., Chakera, J. A., & Gupta, P. D. (2017). Chapter 14—Materials under intense laser irradiation. In A. K. Tyagi & S. Banerjee (Eds.), *Materials Under Extreme Conditions* (pp.

501–532). Elsevier. https://doi.org/10.1016/B978-0-12-801300-7.00014-0 (accessed on 30th July 2025).

100. Nunes, M. d'Ávila. (2014). *Hadron Therapy Physics and Simulations*. Springer. https://doi.org/10.1007/978-1-4614-8899-6 (accessed on 30th July 2025).
101. O'Connor, M. K., & Kemp, B. J. (2006). Single-photon emission computed tomography/computed tomography: Basic instrumentation and innovations. *Seminars in Nuclear Medicine, 36*(4), 258–266. https://doi.org/10.1053/j.semnuclmed.2006.05.005 (accessed on 30th July 2025).
102. Omondi, C., et al. (Kenya). (2021). Characterization of radiation protection X-ray beam for secondary standard dosimetry laboratory at Kenya Bureau of Standards (No. INIS-KE–088). *Kenyatta University P.O. Box 43844–00100, Nairobi (Kenya)*. https://inis.iaea.org/records/2eg71-4m696 (accessed on 30th July 2025).
103. Patel, F. D., Negi, P. S., Sharma, S. C., Kapoor, R., Singh, D. P., & Ghoshal, S. (1998). Dose rate correction in medium dose rate brachytherapy for carcinoma cervix. *Radiotherapy and Oncology, 49*(3), 317–323. https://doi.org/10.1016/S0167-8140(98)00100-5 (accessed on 30th July 2025).
104. Patel, N. S., Fan, P., Chiu-Tsao, S. T., Ravi, K., Sherman, W., Quon, H., Pisch, J., Tsao, H. S., & Harrison, L. B. (2001). Ytterbium-169: A promising new radionuclide for intravascular brachytherapy. *Cardiovascular Radiation Medicine, 2*(3), 173–180. https://doi.org/10.1016/S1522-1865(01)00085-3 (accessed on 30th July 2025).
105. Pauli, W. (1994). Exclusion principle and quantum mechanics. In W. Pauli, C. P. Enz, & K. von Meyenn (Eds.), *Writings on Physics and Philosophy* (pp. 165–181). Springer. https://doi.org/10.1007/978-3-662-02994-7_20 (accessed on 30th July 2025).
106. Pelt, W. R. V. (1999). Ionizing radiation. In *Applications and Computational Elements of Industrial Hygiene*. CRC Press.
107. Podgoršak, E. B. (2014). Interaction of charged particles with matter. In E. B. Podgorsak, *Compendium to Radiation Physics for Medical Physicists* (pp. 299–386). Springer Berlin Heidelberg. https://doi.org/10.1007/978-3-642-20186-8_6 (accessed on 30th July 2025).
108. Podgoršak, E. B. (2016). Coulomb scattering. In E. B. Podgorsak (Ed.), *Radiation Physics for Medical Physicists* (pp. 79–142). Springer International Publishing. https://doi.org/10.1007/978-3-319-25382-4_2 (accessed on 30th July 2025).
109. Poston, D. I., Gibson, M. A., Godfroy, T., & McClure, P. R. (2020). KRUSTY reactor design. *Nuclear Technology, 206*(sup1), 3. https://doi.or

g/10.1080/00295450.2020.1725382 (accessed on 30th July 2025).

110. Prasad, S. G., Parthasaradhi, K., Bloomer, W. D., Al-Najjar, W. H., McMahon, J., & Thomson, O. (1998). Aluminum, copper, tin and lead as shielding materials in the treatment of cancer with high-energy electrons. *Radiation Physics and Chemistry, 53*(4), 361–366. https://doi.org/10.1016/S0969-806X(98)00130-3 (accessed on 30th July 2025).

111. Prokop, M. (2003). Multislice CT: Technical principles and future trends. *European Radiology, 13*(S5), 3–13. https://doi.org/10.1007/s00330-003-2178-z (accessed on 30th July 2025).

112. Pynn, R. (2009). Neutron scattering—A non-destructive microscope for seeing inside matter. In L. Liang, R. Rinaldi, & H. Schober (Eds.), *Neutron Applications in Earth, Energy and Environmental Sciences* (pp. 15–36). Springer US. https://doi.org/10.1007/978-0-387-09416-8_2 (accessed on 30th July 2025).

113. Rabir, M. H., Ismail, A. F., & Yahya, M. S. (2021). Review of the microheterogeneous thoria-urania fuel for micro-sized high temperature reactors. *International Journal of Energy Research, 45*(8), 11440–11458. https://doi.org/10.1002/er.5923 (accessed on 30th July 2025).

114. Rachkov, V. I., Kalyakin, S. G., Kukharchuk, O. F., Orlov, Yu. I., & Sorokin, A. P. (2014). From the first nuclear power plant to fourth-generation nuclear power installations [on the 60th anniversary of the world's first nuclear power plant]. *Thermal Engineering, 61*(5), 327–336. https://doi.org/10.1134/S0040601514050073 (accessed on 30th July 2025).

115. Ritt, P., Sanders, J., & Kuwert, T. (2014). SPECT/CT technology. *Clinical and Translational Imaging, 2*(6), 445–457. https://doi.org/10.1007/s40336-014-0086-7 (accessed on 30th July 2025).

116. Rodríguez-Penalonga, L., & Moratilla Soria, B. Y. (2017). A review of the nuclear fuel cycle strategies and the spent nuclear fuel management technologies. *Energies, 10*(8), Article 8. https://doi.org/10.3390/en10081235 (accessed on 30th July 2025).

117. Rogers, D. W. O., & Townson, R. W. (2019). On calculating kerma, collision kerma and radiative yields. *Medical Physics, 46*(11), 5173–5184. https://doi.org/10.1002/mp.13744 (accessed on 30th July 2025).

118. Roncali, E., & Cherry, S. R. (2011). Application of silicon photomultipliers to positron emission tomography. *Annals of Biomedical Engineering, 39*(4), 1358–1377. https://doi.org/10.1007/s10439-011-0266-9 (accessed on 30th July 2025).

119. Saha, G. B. (2018). Radiation regulations, protection, and uses. In G. B. Saha (Ed.), *Fundamentals of Nuclear Pharmacy* (pp. 219–260). Springer

International Publishing. https://doi.org/10.1007/978-3-319-57580-3_11 (accessed on 30th July 2025).

120. Salem, R., Thurston, K. G., Carr, B. I., Goin, J. E., & Geschwind, J. F. H. (2002). Yttrium-90 microspheres: Radiation therapy for unresectable liver cancer. *Journal of Vascular and Interventional Radiology, 13*(9, Part 2), S223–S229. https://doi.org/10.1016/S1051-0443(07)61790-4 (accessed on 30th July 2025).
121. Sanders, C. L. (2017). *Radiobiology and Radiation Hormesis*. Springer International Publishing. https://doi.org/10.1007/978-3-319-56372-5 (accessed on 30th July 2025).
122. Schardt, D., Elsässer, T., & Schulz-Ertner, D. (2010). Heavy-ion tumor therapy: Physical and radiobiological benefits. *Reviews of Modern Physics, 82*(1), 383–425. https://doi.org/10.1103/RevModPhys.82.383 (accessed on 30th July 2025).
123. Scheck, F. (1978). Muon physics. *Physics Reports, 44*(4), 187–248. https://doi.org/10.1016/0370-1573(78)90014-5 (accessed on 30th July 2025).
124. Schmidt, G. P., Haug, A. R., Schoenberg, S. O., & Reiser, M. F. (2006). Whole-body MRI and PET-CT in the management of cancer patients. *European Radiology, 16*(6), 1216–1225. https://doi.org/10.1007/s00330-006-0183-8 (accessed on 30th July 2025).
125. Schreiber, R. (2009). Pressurized water reactors (PWRs). In *Nuclear Engineering Handbook*. CRC Press.
126. Schumacher, M. (2005). Polarizability of the nucleon and Compton scattering. *Progress in Particle and Nuclear Physics, 55*(2), 567–646. https://doi.org/10.1016/j.ppnp.2005.01.033 (accessed on 30th July 2025).
127. Scott, H. A. (2005). Non-LTE radiation transport in high radiation plasmas. *High Energy Density Physics, 1*(1), 31–41. https://doi.org/10.1016/j.hedp.2005.08.002 (accessed on 30th July 2025).
128. Sharma, S. D. (2024). Radiation environment in medical facilities. In D. K. Aswal (Ed.), *Handbook on Radiation Environment, Volume 2* (pp. 303–345). Springer Nature Singapore. https://doi.org/10.1007/978-981-97-2799-5_11 (accessed on 30th July 2025).
129. Singh, A. K., Siingh, D., & Singh, R. P. (2010). Space weather: Physics, effects and predictability. *Surveys in Geophysics, 31*(6), 581–638. https://doi.org/10.1007/s10712-010-9103-1 (accessed on 30th July 2025).
130. Sizmann, R. (1978). The effect of radiation upon diffusion in metals. *Journal of Nuclear Materials, 69, 70*, 386–412. https://doi.org/10.1016/0022-3115(78)90256-8 (accessed on 30th July 2025).

131. Slomka, P. J., Berman, D. S., & Germano, G. (2014). New cardiac cameras: Single-photon emission CT and PET. *Seminars in Nuclear Medicine, 44*(4), 232–251. https://doi.org/10.1053/j.semnuclmed.2014.04.003 (accessed on 30th July 2025).

132. Smith, J. C. (2022). Chapter 6—Steam generators. In J. R. Riznic (Ed.), *Pressurized Heavy Water Reactors* (Vol. 7, pp. 229–273). Elsevier. https://doi.org/10.1016/B978-0-12-822054-2.00019-2 (accessed on 30th July 2025).

133. Solomon, B. D. (1984). Decommissioning nuclear power plants. In *Nuclear Power*. Routledge.

134. Stabin, M. G. (Ed.). (2003). Radiation protection practice/evaluation. In *Radiation Protection and Dosimetry* (pp. 244–308). Springer New York. https://doi.org/10.1007/978-0-387-49983-3_11 (accessed on 30th July 2025).

135. Stambulchik, E., & Maron, Y. (2006). A study of ion-dynamics and correlation effects for spectral line broadening in plasma: K-shell lines. *Journal of Quantitative Spectroscopy and Radiative Transfer, 99*(1), 730–749. https://doi.org/10.1016/j.jqsrt.2005.05.058 (accessed on 30th July 2025).

136. Stewart, F. A., Akleyev, A. V., Hauer-Jensen, M., Hendry, J. H., Kleiman, N. J., MacVittie, T. J., Aleman, B. M., Edgar, A. B., Mabuchi, K., Muirhead, C. R., Shore, R. E., & Wallace, W. H. (2012). ICRP Publication 118: ICRP statement on tissue reactions and early and late effects of radiation in normal tissues and organs–Threshold doses for tissue reactions in a radiation protection context. *Annals of the ICRP, 41*(1, 2), 1–322. https://doi.org/10.1016/j.icrp.2012.02.001 (accessed on 30th July 2025).

137. Stoller, R. E., & Greenwood, L. R. (1999). Subcascade formation in displacement cascade simulations: Implications for fusion reactor materials. *Journal of Nuclear Materials, 271, 272*, 57–62. https://doi.org/10.1016/S0022-3115(98)00730-2 (accessed on 30th July 2025).

138. Taïeb, D., Hicks, R. J., Hindié, E., Guillet, B. A., Avram, A., Ghedini, P., Timmers, H. J., Scott, A. T., Elojeimy, S., Rubello, D., Virgolini, I. J., Fanti, S., Balogova, S., Pandit-Taskar, N., & Pacak, K. (2019). European Association of Nuclear Medicine practice guideline/Society of Nuclear Medicine and Molecular Imaging procedure standard 2019 for radionuclide imaging of phaeochromocytoma and paraganglioma. *European Journal of Nuclear Medicine and Molecular Imaging, 46*(10), 2112–2137. https://doi.org/10.1007/s00259-019-04398-1 (accessed on 30th July 2025).

139. Ter-Pogossian, M. M. (1977). Basic principles of computed axial

tomography. *Seminars in Nuclear Medicine, 7*(2), 109–127. https://doi.org/10.1016/S0001-2998(77)80013-5 (accessed on 30th July 2025).

140. Theodoropoulos, A. S., Gkiozos, I., Kontopyrgias, G., Charpidou, A., Kotteas, E., Kyrgias, G., & Tolia, M. (2020). Modern radiopharmaceuticals for lung cancer imaging with positron emission tomography/computed tomography scan: A systematic review. *SAGE Open Medicine, 8*, 2050312120961594. https://doi.org/10.1177/2050312120961594 (accessed on 30th July 2025).
141. Theriault, K. (2009). Boiler water reactors (BWRs). In *Nuclear Engineering Handbook*. CRC Press.
142. Theriault, K. (2016). Boiling water reactors. In *Nuclear Engineering Handbook* (2nd ed.). CRC Press.
143. Tomascak, P. B., Magna, T., & Dohmen, R. (2016). *Advances in Lithium Isotope Geochemistry*. Springer International Publishing. https://doi.org/10.1007/978-3-319-01430-2 (accessed on 30th July 2025).
144. Vallabhajosula, S. (2023). Radioactivity. In S. Vallabhajosula (Ed.), *Molecular Imaging and Targeted Therapy: Radiopharmaceuticals and Clinical Applications* (pp. 49–62). Springer International Publishing. https://doi.org/10.1007/978-3-031-23205-3_4 (accessed on 30th July 2025).
145. Verma, H. R. (Ed.). (2007). Nuclear reaction analysis and particle-induced gamma-ray emission. In *Atomic and Nuclear Analytical Methods: XRF, Mössbauer, XPS, NAA and Ion-Beam Spectroscopic Techniques* (pp. 269–293). Springer. https://doi.org/10.1007/978-3-540-30279-7_7 (accessed on 30th July 2025).
146. Vijayalaxmi, Fatahi, M., & Speck, O. (2015). Magnetic resonance imaging (MRI): A review of genetic damage investigations. *Mutation Research/Reviews in Mutation Research, 764*, 51–63. https://doi.org/10.1016/j.mrrev.2015.02.002 (accessed on 30th July 2025).
147. Von Hirschhausen, C. (2018). German energy and climate policies: A historical overview. In C. Von Hirschhausen, C. Gerbaulet, C. Kemfert, C. Lorenz, & P. Y. Oei (Eds.), *Energiewende "Made in Germany"* (pp. 17–44). Springer International Publishing. https://doi.org/10.1007/978-3-319-95126-3_2 (accessed on 30th July 2025).
148. Voyiadjis, G. Z., & Abed, F. H. (2005). Microstructural based models for bcc and fcc metals with temperature and strain rate dependency. *Mechanics of Materials, 37*(2–3), 355–378. https://doi.org/10.1016/j.mechmat.2004.02.003 (accessed on 30th July 2025).
149. Wang, C., Xu, J., Zhang, Y., & Nie, G. (2023). Emerging nanotechnological

approaches to regulating tumor vasculature for cancer therapy. *Journal of Controlled Release, 362*, 647–666. https://doi.org/10.1016/j.jconrel.2023.09.017 (accessed on 30th July 2025).

150. Weil, J. C. (1988). Plume rise. In A. Venkatram & J. C. Wyngaard (Eds.), *Lectures on Air Pollution Modeling* (pp. 119–166). American Meteorological Society. https://doi.org/10.1007/978-1-935704-16-4_4 (accessed on 30th July 2025).
151. Wieler, R. (2018). Cosmogenic nuclides. In *Encyclopedia of Geochemistry* (pp. 317–325). Springer, Cham. https://doi.org/10.1007/978-3-319-39312-4_332 (accessed on 30th July 2025).
152. Yasoob A., N., Abdalameer, N. Kh., & Mohammed, A. Q. (2022). Plasma production and applications: A review. *International Journal of Nanoscience, 21*(06), 2230003. https://doi.org/10.1142/S0219581X22300036 (accessed on 30th July 2025).
153. Yee, E., & Biltoft, C. A. (2004). Concentration fluctuation measurements in a plume dispersing through a regular array of obstacles. *Boundary-Layer Meteorology, 111*(3), 363–415. https://doi.org/10.1023/B:BOUN.0000016496.83909.ee (accessed on 30th July 2025).
154. Yuan, D., & Liu, Q. (2022). Photon energy and photon behavior discussions. *Energy Reports, 8*, 22–42. https://doi.org/10.1016/j.egyr.2021.11.034 (accessed on 30th July 2025).
155. Zaidi, L., Belgaid, M., Taskaev, S., & Khelifi, R. (2018). Beam shaping assembly design of ^{7}Li(p,n)^{7}Be neutron source for boron neutron capture therapy of deep-seated tumor. *Applied Radiation and Isotopes, 139*, 316–324. https://doi.org/10.1016/j.apradiso.2018.05.029 (accessed on 30th July 2025).
156. Zou, J., Guo, D. Q., & Tong, L. L. (2015). Evaluation of RCS injection strategy by normal residual heat removal system in severe accident management. *Annals of Nuclear Energy, 85*, 166–174. https://doi.org/10.1016/j.anucene.2015.05.013 (accessed on 30th July 2025).
157. Zuidersma, W., du Toit, M. H., & van Eldik, M. (2024). Worst-case accident analysis of accident tolerant fuel in NuScale using RELAP5/MOD3-based code. *International Journal of Energy Research, 2024*(1), 8410668. https://doi.org/10.1155/er/8410668 (accessed on 30th July 2025).

INDEX

D

E

F

G

H

I

K

L

M

N

S

T

U